SCOUTING AND PATROLLING

SCOUTING AND PATROLLING

Ground Reconnaissance Principles And Training

by

Lt. Col. Rex Applegate
(U.S. Army — Ret.)

Also by Rex Applegate:
Bullseyes Don't Shoot Back (with Michael D. Janich)
Close Combat Files of Colonel Rex Applegate (with Maj. Chuck Melson)
An Infantryman's Guide to Combat in Built-Up Areas
Combat Use of the Double-Edged Fighting Knife
Kill or Get Killed
Point Shooting (video)
Shooting for Keeps (video)

Scouting and Patrolling
by Lt. Col. Rex Applegate (U.S.A.–Ret.)

ISBN 13: 978-0-87364-184-5
Printed in the United States of America

Published by Paladin Press, a division of
Paladin Enterprises, Inc.,
P.O. Box 1307
Boulder, Colorado 80306 USA
+1.303.443.7250

Direct inquiries and/or orders to the above address.

Visit our website at www.paladin-press.com.

TABLE OF CONTENTS

PUBLISHERS' FORWARD

"Although the basic principles of scouting and patrolling are unchanging, it is recognized that this is the era of the electronic battlefield."

Scouting And Patrolling was originally scheduled for publication near the end of World War II. At that time, our combat experience and intelligence data indicated that there was a lack of basic training data and tactical techniques in the Army scouting/patrolling manuals then available. This book was written to amend that situation. But the sudden end of the war, and a predicted loss of interest in military texts caused the publishers to cancel the project.

An examination of the current U.S. Army manual FM 21-75 reveals that some of this material has been incorporated into Army doctrine. However, it is felt that this book contains important information and tactics still not found in any other current manual; hence the justification for publishing this manuscript, 35 years after it was written.

Although the basic principles of scouting and patrolling are unchanging, it is recognized that this is the era of the electronic battlefield. Satellites, lasers, helicopters, all-terrain vehicles, superior night vision sights and devices, advanced aerial photo techniques, and sophisticated ordnance are now commonly in use among major international armed forces of the world. Yet there are still countless small "brushfire" conflicts and military actions occurring throughout the world today that are fought between poorly-equipped ground forces. Such forces especially are in need of the basic tactics as used by Rodgers' Rangers, over 200 years ago.

The information and techniques presented here were developed and taught to intelligence officers at the U.S. Army Intelligence Training Center at Camp Ritchie, Maryland from 1942 to 1945. Colonel Applegate commanded a training section there consisting of 27 officers and ranking enlisted men who were veterans of recent combat in all theatres of war. This elite group of men studied the scouting and patrolling techniques of the enemy, allies, and our own forces, then formulated the instructional courses presented here.

Some of the author's references to weapons and events may seem a bit outdated to the casual reader. But the publishers feel strongly that **Scouting And Patrolling** is a timeless and important work, and that recognition of this fact is long overdue. Any officer responsible for training or leading his men in scouting/patrolling operations should welcome the addition of this book to his library.

The Publishers

INTRODUCTION

"An intelligent man of good physique, who is confident, aggressive, and self-reliant, is the best raw material from which scouts and patrol members are made."

The modern military commander has many aids for the gathering of information about the enemy, such as the photo-interpretation team, the interrogation team, the counter-intelligence detachment, and the usual exchange of information between higher, lower and adjacent headquarters. Even with these outside aids, the commander is still mainly dependent upon his own intelligence and ground reconnaissance agencies. Often they are the only means of obtaining or confirming enemy information. The personnel of these intelligence sections and reconnaissance agencies must be especially trained in their duties, a great portion of which will be scouting, patrolling and observation.

Once in contact with the enemy, No Man's Land, whether it be 100 yards or 100 miles broad, must be kept under continuous observation and control by friendly patrols. Never should a state of exhaustion of troops or command result in neglect of this observation or reconnaissance.

Unfortunately, observation alone is not always enough. Too often a particular area of enemy activity is hidden from even the best of aerial photographs or ground observers. To confirm previous reports or to gain additional information, the scout or patrol must be used. Since enemy information will not always come easily, the commander must have scouts or patrols capable of going out to get it and of bringing it back.

The scout or patrol member must be a specially selected solider who has undergone intensive training before he can properly perform his mission in combat. He cannot be replaced by any basic private should he become a casualty. An intelligent man of good physique, who is confident, aggressive, and self-reliant, is the best raw material from which scouts and patrol members are made. When the unit commander or his S-2 gives such men proper training, their usefulness to him will be greatly increased.

Active training in scouting and patrolling, which can never be acquired entirely from training films or field manuals, should be given in the field under conditions as similar to actual combat as possible. The training program must be specific and realistic. Only those men with combat experience, who have seen the value of such training, will apply themselves wholeheartedly.

Scouting and patrolling is a subject which even the best men will not learn without a continuous, rigidly supervised training program. Americans are generally too mechanized by machine-age habits to appreciate the necessity of getting down on their bellies and crawling.

It is hoped that this material, which contains training hints and information gathered from a multitude of manuals, combat reports, enemy and allied publications and individual experiences, will be an aid to instructing officers in the training program.

Special training exercises, directed against personnel dressed as the enemy, using enemy organization, tactics and equipment will be discussed later. In some units these facilities may not be available to the training officer, but he can train his men against American troops and organizations with but little change in the discussed training program and achieve the desired results. Basically, scouting and patrolling principles are the same in all armies. Some of the training methods suggested are idealized, such as might be practiced in a special school, but the individual training officer can easily adapt them to his own situation.

Although this book has been written primarily as an aid to training officers of intelligence and reconnaissance units, its use is not limited to them alone. Company grade infantry officers will find this material of special interest, since they are often called upon to furnish and lead patrols for specific intelligence and combat missions.

Enemy and allied armies assign specially trained scouts and observers to their basic infantry units. The Marine Corps has used somewhat similar plans in assigning specially trained groups of scouts to different headquarters as the situation demands. Army directives and memoranda have suggested similar procedure for our infantry units. A special group of selected and trained scouts could be allotted each rifle company and battalion. These units could be used for reconnaissance patrols and as special covering detachments for forward security in approach marches.

A provision whereby these special allotments of trained scouts and observers can be provided to various headquarters is very desirable. Such groups could also be utilized as replacements or used in situations where normal reconnaissance agencies are not sufficiently trained or lack the numerical strength to accomplish the desired missions. Ranger units have been attached to infantry organizations to perform necessary patrol and scouting tasks. Some divisions have organized provisional units locally, calling them "battle patrols." These specially trained units have successfully operated on specific reconnaissance missions because of their selected personnel, aggressiveness and individualized training.

Scouting and patrolling is comparable to rifle markmanship. All members of infantry units must be able to fire a rifle. Similarly, all must be able to perform normal scouting and patrolling missions. Outstanding riflemen are given extra training and equipment and become snipers. In the same way, specially selected and qualified men or units must be trained to perform special reconnaissance and patrolling missions.

This text includes all the source material, instructional methods and references necessary to train individuals and units in foot reconnaissance.

CHAPTER I

INDIVIDUAL TRAINING

"In the military sense, the natural born scout does not exist. He can only be created by proper selection and training."

A military scout must learn to do, through practice and training, the things animals do by instinct. In addition to the animal traits of patience, silent movement, and the instinctive utilization of cover and concealment, the scout must also master many other special skills to enable him to fulfill his combat role.

The word "scout" is in itself an all-inclusive term, and conjures up visions of the skilled plainsmen of our frontier days. But to designate a soldier as a scout does not automatically endow him with all of the desirable attributes of our Indian-fighting ancestors. Because we are decades removed from our pioneer days and because our armies are largely recruited from thickly populated areas, we must place special emphasis upon the selection and training of the military scout. This training must be above and beyond that received in basic military training, in order to produce the ideal reconnaissance scout, observer or patrol member such as manuals describe.

The military importance of the scout in modern war is such that all the special attributes demanded of him no longer come from his environment alone, but can only be acquired by careful selection and training. The raw material of a machine age from which good scouts must be developed is plentiful, but it must be carefully selected and properly trained. The veneer of civilization has practically eliminated the inherent characteristics of our pioneer ancestors which are so desirable. In other words, we must start from "scratch" in the selection of the reconnaissance scout. Beyond the basic qualifications of good physical condition, unimpaired vision, keen hearing and a high mental standard, the desirable traits to make an ideal scout must be developed by training.

Standards of selection must not be too rigid. Generally, however, an appraisal of the age, civilian background and emotional characteristics of the soldier will help. Often the very young are lacking in judgment and patience, making them too impulsive or too ready for a fight. Older, more mature men who have spent the greater part of their lives outdoors in the fields and forests make much better basic scout material. In addition to being more at home in the field, they are usually more reliable, possess greater patience and exercise better judgment at critical moments. The qualities of leadership and personal confidence are often better developed. This should not be construed to mean that a young soldier or a city-bred soldier will not make a good scout, but normally such men will need more training.

In combat, certain men will stand out as individuals and show other desirable qualities, such as aggressiveness and personal courage. They will often take great personal pride in their fieldcraft and ability to outsmart the enemy, and will desire to operate alone on scouting missions. Men of this type make ideal scouts and should be given every opportunity by the commanding officer to exploit such natural tendencies. Such individuals, however, are not usually present in sufficient numbers to assume the entire reconnaissance burden of modern armies. The old concept of the individual scout operating alone in the face of the enemy depends largely upon the presence of such men in sufficient numbers in all units.

Nowadays, two and three man patrols must assume the responsibility for "sneak" reconnaissance. Recent combat has revealed that for operational, psychological and security reasons the two man scout team or patrol (3 men) will best perform the bulk of sneak reconnaissance missions. Although it is true that most sneak reconnaissance missions will be executed by small patrols, the qualities of the individual scout must still be as high as when he was used singly on missions. Training must be directed toward making him the well schooled, self-sufficient soldier that the name "scout" implies.

After basic infantry training, the scout's initial instruction in scouting and patrolling must be centered around three basic points, *camouflage, movement,* and *sound.*

No matter how thoroughly the scout (or any soldier) is schooled in maps, compass, weapons, observation, etcetera, his value in the field is nullified until his training in the basic principles of movement, sound and camouflage has been completed. If a soldier cannot take advantage of terrain, moving properly, invisibly and quietly, he is not a scout. Regardless of his other special skills, he is not of much value as a reconnaissance agent unless he has mastered himself and his movement over terrain under the eyes of the enemy. He must be able to go out and *get* enemy information and *bring it back.*

The scout must be well-trained in the use of artificial concealment aids such as camouflage suits, nets, paint, and other field expedients, but an understanding of the principles of natural camouflage and the proper use of terrain features and foliage must be his principal means of concealment.

Proper movement in the field not only involves the correct utilization of terrain and camouflage but it also demands silence while moving and at a halt. Silent movement over all types of terrain is essential. Too many soldiers advance headlong into enemy territory without avoiding brush, tile, and various other noise-producing obstacles. The scout must be taught to utilize quiet areas, or to silently cross unavoidable noise-producing obstacles, man-made and natural.

Once enemy contact has been established, daytime missions are usually restricted by enemy observation and counter reconnaissance screens. Consequently, most reconnaissance missions will be undertaken at night when the enemy must rely mainly upon his ears to detect the presence of the scout or patrol. For these reasons, every scout should receive thorough training in silent movement, particularly at night. Each training officer should use a silent movement course in his training program, similar to the one described in the exercise section.

The basic elements of courage, initiative, resourcefulness and perseverance should be part of the makeup of all scouts. All of these, however, will be of little use without the trait of patience. Lack of patience leads to carelessness, which is tantamount to suicide when operating under the eyes of the enemy. Patience is not a product of the machine age, nor is it a trait which is naturally found in too much abundance in the American soldier. It is a quality which should distinguish all intelligence personnel. Reconnaissance demands patience; a scout is often betrayed by a lack of it.

Many reports from various theatres of war emphasize the necessity for patience in all actions, large or small. A vet from WW II's Guadalcanal told a story of a day's combat; not a battle of wits, strength or courage, but a battle pitting the patience of a Japanese against the patience of an American soldier. For several hours each waited for the other to move and disclose his position. The Japanese moved first, and the American fired. He did not miss. From WW II to Vietnam, all too often American soldiers have lost these battles of patience.

If patience is not naturally present in the makeup of the scout it must be developed during the training program. If it cannot be ingrained into an individual during training, the probabilities of the success of that particular soldier as a military scout are greatly lessened.

After the soldier has developed through training the ability to move invisibly and silently in the face of the enemy, he must achieve proficiency in many other things before he can become a scout. Where, how, and what to see must next be covered in his training.

He must be something of an engineer to judge the correct load of bridges and roads; he must know how to relay information by semaphore or wire. His training must include hours of compass work, both day and night; he must be very proficient at map reading and he must know how to make planimetric, panoramic and contoured sketches. His knowledge of booby traps, mines, and demolitions must be superior to that of the ordinary soldier. He must know his own army as well as that of the enemy, both as to organization, equipment and tactics.

It is easy enough to see that the demands on the modern scout are considerably greater than those made upon scouts of the past. Present day warfare has made it mandatory that he be trained not only in Indian lore but also in many of the complexities of modern battle.

When the imposing list of subjects necessary for the scout's military education are surveyed it is evident that the ideal reconnaissance scout cannot be trained "on the side," in addition to his regular duties. His training program is such that he must be taken aside and given special intensive training.

In the unit, the amount of training emphasis put on reconnaissance agencies by the commanding officer will be directly reflected by the success of his reconnaissance elements in combat. Many unit commanders have realized the necessity for the extra training of their scouts, and some have organized provisional reconnaissance groups in addition to their regular assigned personnel to participate

Figure 2. This is deception.

Figure 3. This is cover.

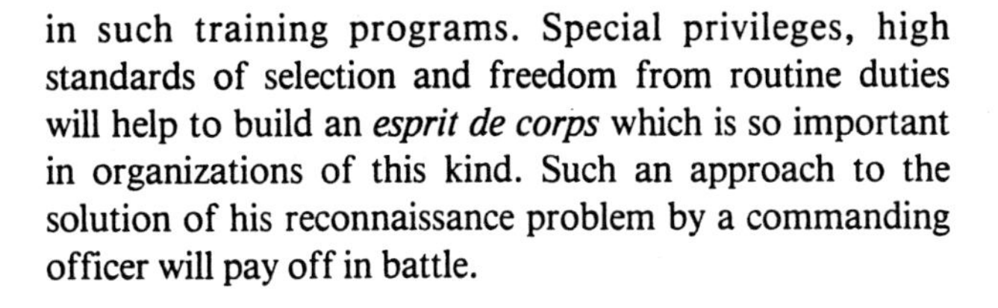

in such training programs. Special privileges, high standards of selection and freedom from routine duties will help to build an *esprit de corps* which is so important in organizations of this kind. Such an approach to the solution of his reconnaissance problem by a commanding officer will pay off in battle.

Night Sight

Scouting and patrolling operations will be mostly undertaken at night, consequently, emphasis on night vision, proper sound identification and use of the ears must be great. There are many men who have poor night sight. This deficiency should be ascertained by tests and it must be corrected or such men should not be used on night missions. An individual with normal eyesight can increase his powers of night vision by training and the use of a few simple precautions.

Generally, night vision is inaccurate because only outlines, not details, can be distinguished. Many times even outlines will be indistinguishable, but regardless of the degree of night vision, the mere fact that the eyes are open means an aid to the individual in maintaining balance, sense of direction, and silent movement.

Consequently, even though sometimes the eyes are not useful in identifying objects they are still an important morale factor. If you put a blindfold over an individual's eyes, he immediately loses confidence and sense of direction. (See blindfold pen in exercise section). When any degree of light is present, however, the proper use of the eyes will greatly increase the effectiveness of the scout. By following proven rules, the value of night sight as a supplement to hearing are greatly increased.

Basically, proper preparation means the conditioning of the eyes by shielding them from bright lights from thirty minutes to an hour before departure on the night mission. The sudden transition of the eyes from bright light to darkness temporarily places the soldier at a disadvantage. In close proximity to the enemy his usefulness may be over before he gains his night vision.

Even if he has gained night vision, any exposure to light, even for a moment, will result in a decrease in his powers of night vision. The more intense the light, the greater the amount of night vision lost. Flares, grenades, mine and artillery shell explosions and the use of flashlights will all affect his degree of vision on the night mission. Unavoidable exposure of the eyes to such conditions should result, if possible, in a period of inactivity while the eyes are readjusting themselves. Fatigue, certain vitamin deficiencies, as well as lack of preparation, will cause an increase in night blindness.

Anyone who has had the experience of going from a brightly lighted street into a darkened theater can appreciate the meaning of the development of night vision.

Figure 4. Everything around you is light or dark, or a tone between these extremes. What shade are you? In medium light, you are a middle tone: "A." In shadow or with light behind you, you appear dark: "B." In direct sunshine you are light in tone: "C."

Figures 5 through 12. Do not be a member of a suicide patrol! In battle, basic principles of camouflage, cover, concealment, and movement must always be observed. Violate them and you die!

The best way to see something at night is to examine it from the corner of the eye. The individual must look at things from the corner of his eye; learning to control his eyes by never looking directly at the object he wishes to see. He must learn that when his eyes are drawn irresistibly to an object – and many times this will happen – he must force them to slide over to the other side, and look again at the object from the other corner of his eye. In night observation, he should never try to sweep the sky, ground or horizon. Objects cannot be easily distinguished while the eyes are moving. Patience becomes a vitally important factor in night observation. Many times a faint object may not be discernible until after many examinations.

Any one who has hunted quail in the early morning or watched deer in the dusk will know that he can look directly at such a naturally camouflaged animal for a long time before he sees it. In darkness such an object is even harder to distinguish, because it cannot be seen if stared at directly. The soldier must look at the object again and again, first one side then the other, until it takes definite shape and is identified by its outline.

Precepts of Better Night Sight

(1) Protect the eyes from strong light.

(2) Use the corners of the eye while observing, moving from point to point in quick jerky movements. Short pauses are better than long, sweeping movements and long pauses.

(3) Be thorough, patient and systematic while observing.

(4) Learn to identify your own and enemy objects by silhouettes because details will not ordinarily be visible.

(5) Don't eat too much before night work.

Sound

The scout must depend on his ears as well as his eyes, especially when he operates at night. He must be able to distinguish and identify various sounds. The common noises of troops in the field should indicate to him their activity. He should be able to distinguish the click of a rifle bolt from the snapping of a twig. The rate of fire, and other distinctive sounds of combat, should indicate to him whether it is his own or the enemy's weapons. The success of his mission will often depend on his skill in interpreting sounds in the field.

In the jungle, where snipers hold their fire until the enemy is within extremely close range, American soldiers have learned to listen for the sound of the enemy operating the bolt of his rifle. For instance, during the WWII Solomon Island Campaign, our soldiers located many Japanese snipers from the sound of their Arisaka rifle bolts snapping shut, enabling our troops to eliminate these enemy snipers.

False sounds can be used to deceive the enemy. Jericho is reputed to have been taken by a few men armed with trumpets. In a like manner, a twig can be broken or other noise can be made to draw enemy fire, thus enabling other observers to locate enemy positions. A loud sound can be used to mask other sounds thus concealing certain activity to the enemy.

Sound can be used to estimate distance. It is a known fact that sound travels at approximately 1080 feet per second. If the flash of a cannon is observed and two seconds elapse between the flash and the report the distance from the cannon to the observer is 720 yards. (Two seconds x 360 yards per second.)

To locate objects, the ear is not as skillful as the eye. The ears, nevertheless, can be very effective in determining the position of sound. If a repeated sound is familiar, so that the man knows its approximate distance, he can quite accurately locate the position of the noise-making object. He will notice sometimes that the sound is louder in one ear than in the other; he should then turn his head so that the sound registers equally in both ears and his nose will point in the general direction of the sound. By stationing two men at different points, using intersection, a position can often be quite accurately determined.

Precepts of Good Hearing

(1) Keep the ears free of wax, else they may be deafened, particularly in hot weather. Wax should be removed by a medical officer whenever possible.

(2) Avoid colds and keep physically fit. If you have a cold, it may be disrupting to subject your ears to a great change of pressure, such as is experienced when taking off and landing in aircraft. A cold is apt to close the tube from your mouth to the middle of the ear, to an extent that pressure on the inside of the eardrum cannot be equalized to the outside pressure. Unequal pressures can deafen.

(3) Avoid exposures to loud sounds, such as heavy gun fire, if you want to be able to hear faint sounds. The effects of loud sounds may easily last for half an hour, sometimes even for a day. Protect your hearing with cotton as you protect your eyes with goggles.

(4) If you do not want to be temporarily deafened by your own rifle fire, plug your left ear with cotton. It is the one that gets the explosion from the muzzle of your gun.

(5) If you are in the midst of loud noises and want to hear speech, plug your ears with ear-plugs, cotton, or even your fingers.

(6) Listen attentively when you know what kind of sound to expect. On patrol and reconnaissance keep quiet and listen.

(7) The steel helmet and the liner definitely limit the hearing ability of the soldier. On patrols and guard duty at night, they should not be worn if the most is to be expected of the ears.

Figure 13. Progressive steps in camouflage.

Figure 14. Camouflage background the same.

Figure 15

Figure 16

Figure 17. Movement, bowel or otherwise, attracts attention.

Figure 18. Don't be an ostrich!

Figure 19. Death.

Figure 20. Kill that shine or it will kill you.

Figure 21. Everything concealed but rifle.

Figure 22. Life.

Figure 23. Use grease, cork, and mud.

Figure 24. Utilize shadows, don't make them.

(8) Get interested in sounds. You can learn to recognize many different noises when you have trained yourself by constant practice. Some men can distinguish the difference between makes of airplanes by the sound of their motors. They cannot describe the difference verbally, but they inherently recognize it. Trained woods and jungle fighters quickly learn the differences between sounds of human and animal movements in the undergrowth, though such noises are similar to the untrained ear.

(9) In cold weather all sounds are amplified and carry much further.

To avoid ear injuries caused by loud sounds the following precautions should be taken:

(1) Use ear plugs or cotton when they will not interfere with the distinguishing of commands, and when those who give commands are using such precautions. Put your fingers in your ears and open your mouth wide when a nearby gun is about to be fired. Our helmets protect the ears from blasts but not from loud sounds. Some helmets actually distort the localization of sounds.

(2) Keep your ears away from areas where the sounds from gun discharges are greatest. Keep away from any position ahead of the gun muzzle. Face the direction from which the sound will come, turning neither ear into it. When you shield one ear, you are generally exposing the other, unless you protect it with the pressure of a finger.

(3) Hunt for objects to screen the ear. Keep behind walls or other protective barriers when an explosion is imminent. Otherwise, lie down in a hollow, because the explosion generally rises from the ground. You are better off in a prone position.

(4) Open your mouth wide when an explosion is due. Doing so helps equalize the pressure on both sides of the eardrum and may save you from having the drum ruptured.

(5) Anticipate the explosion, since, if you are prepared for it, your middle ear muscles set themselves to resist. These muscles are very differently attuned in listening for a faint sound and anticipating a loud one.

Kill or Get Killed

The mission of the individual scout is to get enemy information and to get it back. If he is properly trained and takes the necessary precautions, he normally will not have to fight. However, on those occasions when he is projected into unexpected combat, he must be able to protect himself and the information he has concerning the enemy. By necessity, he carries few weapons, but he should be trained to utilize any weapon which comes readily to hand. He should be able to use enemy equipment as effectively as his own when forced into combat.

The soldier's natural fighting ability should be developed to a point where he can kill quickly and effectively in all close-quarter combat with or without weapons.

Intensive close combat training is a valuable adjunct in the training or reconnaissance personnel. It helps to instill *self-confidence* and an *offensive spirit*, while at the same time it develops the individual's fighting ability with or without weapons. A personal feeling of confidence in his own ability and in training are indispensable prerequisites for any soldier, but it is much more important for the reconnaissance scout to develop such confidence. There is a direct relationship between *close combat training which increases self-confidence and increased aggressiveness by scouts and patrols.*

The average soldier operates against the enemy in battle with other soldiers and group psychology plays a great part in how he conducts himself in battle. The reconnaissance scout, on the other hand, is an individualist who must operate in many situations without moral or physical support from his fellow soldiers. He must have self-confidence as well as personal courage.

Prior to and during any training program for reconnaissance personnel, physical conditioning is of utmost importance. Long hikes, mountain climbing, and other hardening processes in all-weather conditions should have their place in such a training program.

Practical close combat training, rigorous physical conditioning, and continuous field exercises are requisites of all Commando and Ranger units. Consequently, such organizations produce excellent scouts and reconnaissance or combat patrols. The doctrine of the offense must be stressed at all times, and aggressiveness and tenacity in the individual must be developed to a point where they reflect in his performance of combat missions.

Although the psychology of hate and the killer instinct must be fostered in training all soldiers for combat, the reconnaissance scout must never lose sight of his primary mission, the gathering of information. Nevertheless, he must have the ability and the confidence to take care of himself when the fulfillment of his mission is endangered.

The following subjects should be taught in any training course on close combat*:

Unarmed Offense	24 Hrs
Disarming	6 Hrs
Strangulations (Sentry Killing)	4 Hrs
Knife Fighting and Disarming	6 Hrs
Carbine, Pistol, SMG (Combat Firing	24 Hrs —
Total	64 Hrs

* The text material can be found in Rex Applegate's book, *Kill or Get Killed,* available from Paladin Press, Post Office Box 1307, Boulder, Colorado 80306.

Figure 25

The subjects listed and the number of hours suggested are ideal. This instruction can be given at regular periods throughout a training program and can be combined with the general physical conditioning program.

In the conditioning program, it is best to inject as many games and exercises as possible which have the element of body contact. This type of program will help develop aggressiveness and the offensive spirit which is so important. Too much cannot be said about the value of close combat training and physical conditioning.

The use of obstacle courses, stalking courses, battle courses and other types of conditioning must be continuous. If the scouting is to be done in the mountains, use the mountains for conditioning. The value in a program of this kind is most apparent when it is continuous. A gradual hardening process then takes place which will pay dividends in combat, for no matter how much physical punishment the troops take in maneuvers and training, they will take still more in combat.

A study of the racial and fighting characteristics of the enemy and his fighting methods will add value and interest to training. The methods of operating enemy small arms, grenades and munitions should be covered thoroughly. If available, captured enemy weapons should be studied and used for instruction. Emphasis should be placed on the loading and firing of these enemy weapons instead of the field stripping, nomenclature approach.

The remaining chapters of this book completely cover subjects which should receive the major training emphasis. However, any local training program must include emphasis on the following subjects.

(1) Mines and booby traps and their tactical employment (own and enemy).

(2) Sniping and its tactical employment.

(3) Means of communication (visual, voice and message writing).

(4) Organization, equipment and tactical principles of the U.S. Army, including capabilities and limitations of the various weapons and an emphasis on intelligence agencies.

(5) Organization, tactics and equipment of the enemy army.

CHAPTER II

TERRAIN APPRECIATION

"To understand the terrain is to know what the enemy can plan."

Terrain, not any extraordinary enemy tactics, is usually the determining factor in battle strategy. Terrain determines our tactics and it molds the enemy's tactics as well. If a soldier can analyze the terrain, he can analyze the possible action of the enemy.

A real appreciation of terrain and its influence on the character of operations is of prime importance to the individual scout or patrol leader. Without it, he cannot safely and intelligently perform his mission. With it, his chances of success are multiplied many times. All reconnaissance personnel should be well versed in the principles of terrain analysis.

Instructors and students in scouting, patrolling and observation should concentrate on the principles of terrain appreciation as they can be applied to the immediate problems confronting the S-2 or unit commander in his individual sector. No attempt should be made to enter into the grand tactical and strategical picture. The prime purpose of instruction in terrain appreciation of this type is to teach the evaluation of the area concerned from the viewpoint of local reconnaissance operations. Secondly, it is to determine the effect of terrain on the lines of action open to the enemy.

The topographical character of an area often exercises a decisive influence on the course of a battle. Through an analysis of the terrain the course of local military operations can be evolved, and, concurrently, lines of action open to the enemy can be partially determined. Any study of military history will bring out examples in which the commander, after learning of the terrain from ground reconnaissance, was able to utilize his knowledge to defeat the enemy.

Well-known to the American schoolboy is the Battle of the Heights of Abraham, at Quebec, during the French and Indian War. It was here that General Wolfe, commanding the British forces, was able to score a decisive victory over the forces of the French general, Montcalm. The British forces had twice attacked the citadel of Quebec and were repulsed. The French were in a strong position on the Heights which were over 300 feet above the banks of the river at their base. Heavy losses had occurred in frontal attacks against this terrain obstacle. Further upstream an unguarded trail was discovered leading to the top of the Heights. After a diverting action, Wolfe was able to send a large part of his forces up the trail and the French were met on even terms on the Heights and defeated. Although the chance Wolfe took was hazardous, he had a better knowledge of terrain than his enemy and victory was won.

In our WW II Tunisian campaign, our troops won a decisive victory at El Guettar by outflanking strong enemy defenses. Reconnaissance and knowledge of the area procured from local inhabitants enabled American troops to cross undefended mountain terrain, considered impassable by German and Italian defenders, so that the American flanking action directly resulted in a swift victory. Thousands of casualites were saved when the position fell without the necessity of a frontal assault.

Figure 26

Figure 27

It is not enough to study a hostile area and pick out an ideal route through which to perform reconnaissance in the area. Consideration must be given to what the enemy will do to prevent such reconnaissance from being effective. The limitations and capabilities of various infantry arms on the terrain must be known. Comparable weapons of all armies are relatively the same. Basically, the principles of their use are alike. If a scout, for example, knows the general tactical employment of an American mortar, he will be better equipped to locate enemy mortar installations. In other words, if he knows where various weapons would be used taking advantage of the terrain by our army, he can reasonably expect that the enemy will make the same general use of their weapons. This automatically limits the scope of his reconnaissance and it becomes that much easier.

Evaluation of the Terrain

No matter what type of ground and no matter what the tactical situation, terrain can always be evaluated in terms of five factors: (1) Observation, (2) Field of Fire, (3) Concealment and Cover, (4) Obstacles, (5) Communications.

Observation is protection against surprise. Observation of the ground on which a fight is taking place is essential in order to bring effective fire to bear upon the enemy. Observation affords information on what the enemy and the commander's own troops are doing and makes it possible for the commander to control operations of his own troops. The value of cover and concealment is based on denial of observation to the enemy. Defilade areas free from observation should be used by the scout.

Fields of fire are essential to defense. A good field of fire is prerequisite for the most effective employment of firearms. The best field of fire is that over level or uniformly sloping open terrain - terrain over which the enemy can be seen, and over which he has little or no protection from the effective range of infantry weapons. Upon the scout's knowledge of enemy weapons and their identification by sound, and their use in respect to terrain limitations, depends the success of the mission. The perfect field of fire for all weapons is rarely obtainable. The scout must be able to recognize readily the best possible fields of fire for the enemy weapons and those zones where the enemy is limited in his coverage by terrain irregularities. Once he recognizes these danger areas he can plan his actions accordingly.

Concealment means invisibility from the air and ground and may afford protection only when the enemy does not know that the natural or artificial feature is occupied. Cover includes both concealment and the protection from fire provided by favorable peculiarities of the terrain, natural or artificial. The ideal position for defense provides concealment and cover for the defenders with neither cover nor concealment in front to aid an attacking enemy. Attack is favored by terrain that offers good concealment or cover to approach the enemy and it is

Figure 28

this type of terrain that favors the approach of the scout or patrol.

Obstacles are of special importance in modern warfare because of mechanized units. In general, they are defined as effective obstructions to any military force. Some of the natural military obstacles are mountains, rivers, streams, bodies of water, marshes, steep inclines, and heavily wooded terrain. Mountains which are parallel to the direction of advance limit or prohibit lateral movement and protect the flank. When perpendicular to the advance they are an obstacle to the attacker and an aid to the defender. Rivers are similar to mountains in their effect on forces moving parallel or perpendicular to them. In addition, rivers flowing parallel to the advance may be used as routes of supply. Marshes frequently provide more delay to an advance than do bodies of water, because it is generally more difficult to build causeways than bridges. Mechanized vechicles can be restricted in movement by dense woods, marches, steep inclines, gullies, stumps, large rocks, and bodies of water three or more feet in depth. Consciousness of these obstacles and their effect on military action makes the scout better able to act as the "eyes" of his commander.

Routes of communication allow the movement of troops and supplies to the front. These communication routes are roads, railroads, and waterways. Because they are indispensable to units in both defense and offense, they are of particular concern to the terrain analyst.

APPRECIATION BY SCOUTS AND PATROLS

Because terrain appreciation is nowhere more important than to the patrol, an efficient and successful patrol must master the terrain before and during the operation. Patrol·leaders, patrol members and scouts must know the irregularities of the ground and understand how it effects and limits their operations.

The scout who seeks out the enemy must know where to look for him, where to expect him. He must know what terrain objectives attacked by ground forces are usually located near the enemy artillery area. One objective may be a terrain feature affording commanding observation; another, a critical point in the hostile command system or an essential supply route, and a third, an obstacle to armored forces.

The scout must know the military value of terrain compartments especially in regard to their influence upon the disposition of the enemy. When the terrain features inclosing the area prevent direct fire and ground observation into the area from positions outside, the area is called a compartment. Upon its location, size, and shape depend its suitability for tactical use. The field manual says: "A compartment whose longer axis extends in the directions of movement of a force or leads toward or into a position is called a 'corridor'; while compartments extending across or oblique to the direction of movement of the force or its front are designated as cross or lateral, or oblique, compartments." In general, a corridor favors

Figure 29

the attacker, and a cross compartment favors the defender. Therefore, as far as the scout is concerned, an understanding of our tactical principles can be put to best use in applying this knowledge to determining the enemy's lines of action. Tactical knowledge with this end in view must be emphasized.

Every scout should know the principles of infantry weapon employment. If he knows about the American principles of employment, he will be better equipped to locate comparable enemy dispositions. He will look for mortars in a defiladed area and on the reverse slope of a hill. By knowing this in advance, his scope of reconnaissance will thus be limited, and his reconnaissance made that much quicker and easier. Needless waste of time searching for mortar positions on forward slopes of hills or on flat terrain is automatically eliminated.

The scout should know how mortars, mines and light and heavy automatic weapons are employed. If he runs across one machine gun, he should know that there is in all probability another supporting machine gun not far off. He will expect to find machine guns in partial defilade, though not necessarily. He will realize that terrain commading the flanks will likely hold machine guns, since flanking fire is a cardinal tenet in machine gun employment. That terrain facilitating the use of grazing fire is an obvious indication of a machine gun set-up and that interlocking fire is basic to all defensive positions must always be kept in mind.

If the scout knows fundamental tactics, he knows that antitank guns are placed at the heads of avenues of approach and to cover road junctions. He will look for them on the enemy's front lines, as well as on the first terrain feature back. When he locates a mine field, he can expect it to be covered by anti-tank guns and machine guns and search for them accordingly. It is the scout, sound in the knowledge of tactical principles, who brings back accurate enemy information satisfactorily - and safely, too, for he is not likely to blunder into obvious fields of fire or enemy mine fields.

Patrol leaders and scouts must know the terrain in order to select good routes. A preliminary survey of the terrain to be reconnoitered must be made. Once a route is selected for a daylight patrol, the use of this terrain during the mission means not only concealment, but hours of hard work in the daylight in selection of routes which have easily recognized landmarks. Prominent terrain features, power lines, walls, houses and roads picked out in daylight can serve to guide the patrol at night. Yet, it must be remembered that most terrain features look vastly different at night from what they do in the day. This must be borne in mind whenever making a daylight reconnaissance prior to a night patrol - and making such a reconnaissance or making a thorough terrain study by use of aerial photos is a *must* before all night patrol actions. Maps and photos can be taken on the patrol mission as a means of orientation if desired. The making of marks and

Keep away from the other kind of NIGHT SKYLINE where moonlight and shadow meet.
Keep right in the Shadow.
Figure 30

Figure 31

notations on these maps and photos that will violate security in case of capture should be avoided.

Situation permitting, the patrol leader should divide the area to be covered on the mission into sectors. Near his own lines at a point of forced passage, or where the first elements of enemy organizatons have been recognized, the patrol can travel quickly. The closer he gets to the enemy, the slower he must proceed. Officers who overlook this fact in assigning patrol missions find that a green patrol wastes too much time in a safer zone. As a result they are late in coming back, or hurry in the danger zone to make up time – in this way betraying their presence to the enemy. Scouts and patrols are too often expected to bring back enemy information in thirty minutes when actually hours are necessary because of the difficult terrain and dispositions of the enemy. *Time analysis of the mission is of major importance.* Scouts and patrols must be allocated enough time by the dispatching officer to prepare for and complete the mission.

The following zones are generally recognizable in the terrain after study in the sector where the patrol is to operate:

1. A *safe zone* is usually removed from enemy observation where men can move quickly and in loose formation near their own lines.

2. An *alert zone* where more cautious tactics are required, but speed is to be maintained whenever possible.

3. A *zone of approach* where the patrol must work cautiously, creeping, crawling, crouching, moving by sound bounds, always ready for the ambush and combat. Ambush points, especially in the case of night patrols, can be fixed by knowning enemy habits, terrain, and the time required to reach a certain point from the point of departure. These must be explored by scouts before committing the patrol.

TERRAIN FEATURES

It is very necessary that there be standardization of terms describing terrain features. The description of terrain features and the recognition of these ground forms when they are named must be such that there is no lack of understanding between the message writer and recipient as to their meaning.

Actual classroom instruction, followed by outdoor exercises is necessary to enable scouts and patrol members to describe and report terrain features accurately. The following standard definitions of terrain features may be used:

1. *Valley* – Tract of land situated between ranges of hills, ridges of mountains. May be traversed by a river. This feature is sometimes referred to tactically as a "compartment" when the features enclosing the valley (ridges) prevent direct fire and ground observation into the area from positions outside. Terrain compartments are classified in accordance with the direction of the larger dimension in relation to the unit utilizing the terrain. Thus a compartment whose longer axis extends in the direction of movement of a force or leads toward or into a position is called a "corridor"; wide compartments extending across or oblique to the direction of movement of the force or its front are designated as cross, lateral, or oblique compartments.

2. *Clearing* – an open spot cleared of timber or jungle. It will usually be different or unique in appearance, compared with surrounding woods.

3. *Gentle Slope* – A gradual incline or slant upwards or downwards.

4. *Cut* – A surface excavation made in the construction of a roadbed or path.

5. *Peak* – The sharp pointed summit or jutting part of a mountain or hill; greatest height; apex.

6. *Draw* – A natural fold in the ground between two small hills, low-lying ground running between two hills or ridges. A small valley or drainage line.

7. *Road Center* – The middle of a road.

8. *Saddle* – A depression across the summit of a ridge. A low depression along a ridge line or between two adjacent hills.

9. *Ravine* – A long deep hollow worn by action of a stream.

10. *Crest of a Ridge* – The top edge of a ridge similar to the top of an animal's back.

11. *Road Fork* – The branch caused by a junction of two roads.

12. *Skyline* – The line where the land appears to meet the sky.

13. *Cliff* – A sheer, high rock or bank. A precipice that may overlook water.

14. *Bluff* – A high, steep bank, usually with flattened front rising steeply and boldly.

15. *Abrupt Slope* – The steep incline of a hill, road or bank.

16. *Culvert* – A closed channel, such as a pipe, used to carry drainage under and across a road or railroad.

A large chart should be prepared which shows a landscape with the following terrain features: draw, clearing, crossroads, bluff or cliff, road fork, gentle slope, fill, military crest, ravine, cut, valley, saddle, skyline, topographical crest, road junction, peak, steep slope, ridge, hollow and culvert. Each one of these features should be numbered and the key printed on the side of the chart. This key should be covered with a paper strip.

The instructor, after explaining the necessity for standardization of terms describing terrain features and covering the above material by using a chart, can conduct a short test. This can be done by having the student write the correct name as the instructor points to each numbered terrain feature on the chart. After the test, the instructor removes the strip of paper from the key and the students correct their papers.

The students should then be taken to a previously selected piece of terrain where all of the above features are visible. Here the instructor should call on various students

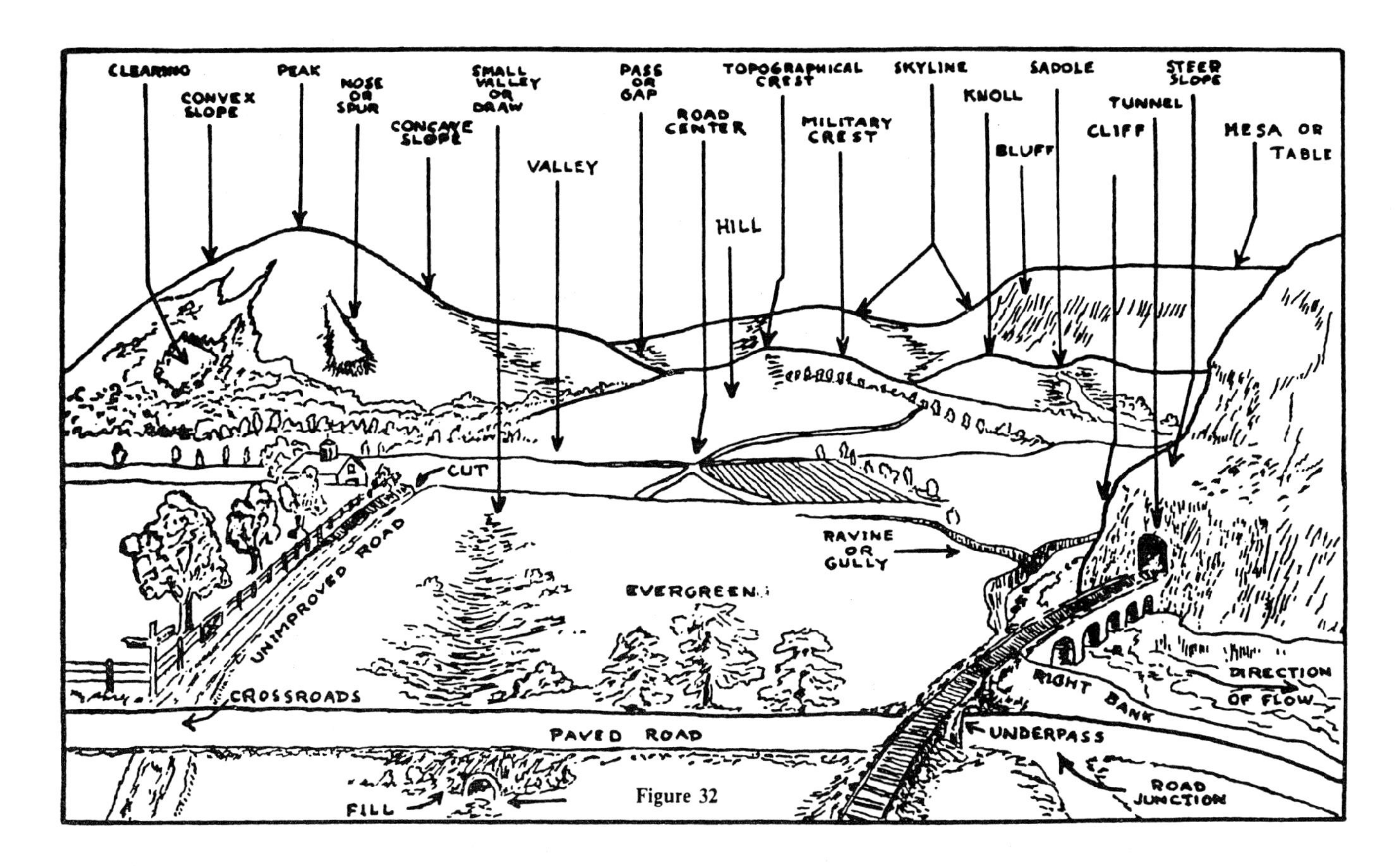

Figure 32

to point out the terrain features which have been previously studied and discussed in the class room.

AERIAL PHOTOGRAPHY

In recent years the interpretation of aerial photos, both vertical and oblique, has become so accurate that capabilities of the enemy can be deduced with uncanny precision. Intelligence of the enemy's organization, disposition, activity and equipment which once took a patrolling mission many hours or days to ascertain, can often be gathered in a very few minutes and with a minimum of danger, by air photograhy. From large scale verticals (1/1000), detailed information of enemy gun positions, mine fields, trenches, road blocks, material, reconnaissance of routes, assessment of bomb damage, selection of artillery and bomb targets, measurements of gradients, the spans of rivers, and height of their banks, the rate of flow of the water, the levels of possible airfields, etcetera, can all be determined by rather simple computations and with relatively simple equipment. From small scale verticals (1/16,000) topographic information for existing maps may be derived, and in certain types of terrain these photomaps can be used as map substitutes. With the WW II innovation of the "floating line" method, the relative visibility from different heights can be seen at a glance. This method is used as a reliable means of establishing OP's.

Because all this information can be gathered within so much less time, with so much less danger, and in most cases, with so much more accuracy, there has been a change in some theaters in the function of scouting and patrolling in the sphere of reconnaissance. Rather than argue over the merits of air versus ground reconnaissance, it should be realized that the two complement each other.

There is no doubt about the fact that the development of aerial photography has eliminated the need for much daylight reconnaissance and thus has greatly eliminated the danger of disclosing to the enemy, by intensive patrolling in a certain area, possible plans of impending operations.

Regardless of the accuracy and usefulness of aerial photography there is still no substitute for direct personal observation by trained scouts and patrols. Patrolling will often have to verify things that photos can only suggest and in many cases ground reconnaissance will have to cover all areas concealed from the photo by weather conditions, dense vegetation and camouflage. This is particularly true in jungle areas.

Along with the above mentioned aids to terrain study the use of th oblique aerial photo can be useful as interpreted by unskilled personnel, because it represents the ground from a more familiar point of view than the vertical photograph. It gives some indication of differences in the height of the ground without the use of special equipment. Larger areas can be covered with obliques and at times they are considerably easier to obtain. They are useful for detailed study prior to the mission and can be used to pinpoint definite areas when the scout or patrol is interrogated at the end of the mission.

The vectograph type of aerial photo which gives a three dimensional picture of terrain so that any novice can study it is another useful recon tool. Great use can be made of vectographs on patrol actions, and should be carried on a patrol mission for constant reference in much the same manner as a map.

Previous to the patrol's starting out, selection of a route hidden from enemy observation can be facilitated by application of two methods. The first is that of profiling. This is a means of determining the visibility or the defilade of points or areas from any selected point on a map. In profiling, the ground on the map is constructed so as to form a profile - a cross-section of the earth's surface - and in this way the slope of any line is shown. By picking out possible enemy OP positions and profiling the terrain from that position to our lines, we can determine which routes toward the enemy will be hidden from his view. Even if the exact location of the enemy OP is not known, its probable position on commanding ground can be fairly accurately plotted, and a route selected accordingly. Combat experience has served to point out the fact that prominent terrain features in enemy territory are likely to contain OP's. Profiling is only possible with accurate maps and involves a great deal of tedious work.

A more accurate route selection method relies on the use of aerial photos, and the floating line technique. This method may be used to excellent advantage.

The technique is simple. Select from an aerial photo or ground observation a piece of commanding ground held by the enemy. Locate on this commanding ground probable enemy observation post sites. From these probable locations draw lines toward the position of your unit to corresponding points on both photographs of a stereo pair. (These lines must be carefully and accurately drawn.)

When viewed stereoscopically the line will either float in the air or merge into terrain features and cover. If the line floats, it means that the enemy has observation to that point. If the line seems to penetrate an object, observation beyond that point is masked. By intelligent use of this method, defilade areas may be plotted on a map that are free from enemy observation. This knowledge of defilade areas can be used in determining general routes for reconnaissance missions. It also can be used for picking future observation points for our own troops and to a certain extent it will help to determine fields of fire for flat trajectory weapons.

Although use of either of these methods will show areas concealed from enemy observation and will help plan probable patrol routes it must be remembered that the enemy also knows they are defiladed. Recognizing this it will be farily certain that such areas will be covered by enemy MG gun emplacements or will be probably places of ambush. Points on ground he can see can be covered by artillery or mortar fire. His blind spots he will protect by other means. This knowledge will aid in localizing the area in which to look for his forward dispositions.

The study of terrain must be furthered by use of maps, photos, and actual eye-witness familiarity. All help increase knowledge of the ground. Anything to develop or increase this knowledge should be used if available. The subject is so important to military operations that local inhabitants should be used as guides when security permits.

Terrain appreciation is as important to the scout or patrol leader on his mission as it is to the general in his strategy.

CHAPTER III

OBSERVATION

"The importance of observation superior to that of the enemy cannot be over emphasized."

Observation is a military art which is engendered during war and neglected in peacetime maneuver. The proper selection, manning, and operation of the observation post is all important to the success of battle operations and continuous observation is essential to the success of every unit, whether a rifle platoon or an army corps. The successful commander maintains continuous ground and aerial observation making use of planes, observation posts, and patrols to obtain maximum information covering the movement and disposition of the enemy. Members of ground reconnaissance agencies must be trained in observation and recognize that it is *primary* to the successful fulfillment of their scouting and patrolling missions.

Because good observation is so necessary for success in battle, it must be obtained by any means regardless of the obstacles or cost. If a lone tree or building is the only place from which ground observation can be carried out, it must be used, even though the use of such an OP site is in tactical violation of principles laid down in manuals. The use of aircraft to supplement limited ground observation has greatly alleviated situations where an impossible terrain has limited the use of the OP, but the observation post must be established and manned wherever and whenever the situation and terrain permits. Observation from the ground or air should always be confirmed by ground reconnaissance, if possible. Areas denied or concealed to the observer should be investigated by scouts or patrols.

The OP, no matter how well located, camouflaged or equipped, will not give maximum service if the personnel manning it have not been carefully selected and trained. Reliable and trained men are necessary for the job. Intelligent men with good eyes and ample patience, who will be attentive to details, are the ideal raw material from which trained observers are made.

Under battle conditions the average soldier can not be trusted to give a reliable and accurate report of enemy activity, especially when he has not been trained and tested in observation. In operations, some commanders have accepted at face value the reports of untrained and untested scouts and observers. Many times these reports have decisively influenced their course of action. Serious losses have been incurred in some instances because no attempt was made or could be made to confirm such reports.

The individual tendencies of each man must be considered on scouting and observing missions. If the personal characteristics of the individual are known, the credence of his reports can be better evaluated.

Exercises in observation must be conducted to test the individual's reliability. A man who fails to remember or see the things he should, may elaborate his report to make it look well. Another may use definite terms such as "long", "large", or "body" to describe specific things.

Exactness must be cultivated. Enemy troops should be reported by number instead of using terms such as squad, battery, platoon. In order to get reports as accurate as possible, tests should be improvised so that the accuracy and reliability of each man who will be used as an observer becomes a known factor.

A most important point to emphasize during the training of men in observation is that they should report everything they see or hear. The commander and his staff do the evaluating and interpreting, not the observer. Nothing is too insignificant to higher headquarters; sometimes one small fact may be the final clue to enemy action. For example:

"In 1918 a regimental observation post reported the movement behind the enemy front lines of a few Germans in light gray uniforms with green collars. This meant nothing to the observer, but after filtering back through channels, it was a most vital identification. It confirmed other evidence of an impending German attack. These uniforms indicated a reconnaissance by officers of the 'Jaeger Division' long missing from the front. This was a first class assault division which generally led major attacks." *(S-2 in Action,* by Colonel Shipley Thomas).

In another case various apparently unrelated events, not one of which alone appeared particularly significant to the observer, were tied together by higher headquarters to make a very important discovery involving future enemy action. The following example of unrelated events properly catalogued by an observer affords an interesting example of a deduction of future enemy action by a regimental S-2, from what appeared to be a handful of individually insignificant events. During the last three weeks of July observers reported:

(1) July 7/31, the enemy dropped six to twelve shells on the line near Flirey Railroad Bridge each afternoon.

(2) July 22, enemy planes attacked and destroyed two of the observation balloons.

(3) Week of July 24/31 four German officers were reported observing the lines near the railroad bridge regularly each afternoon.

(4) Night of July 29/30, a patrol learned that the German front line was manned somewhat more heavily than usual, and that they had camouflaged but failed to close a gap in their wire which had been cut by the Americans some time previously.

(5) July 29/30, six enemy planes flew slowly along the entire front.

(6) July 28/30, each morning and afternoon the crossroads to the rear of the American position was shelled by ranging bursts of shrapnel.

From these various observations covering a three week period, the S-2 was able to foresee a large scale German attack on July 31. His reasoning, based on knowledge of previous enemy action, was as follows:

(1) The few shells landing near the R.R. Bridge were to determine and to verify the range, probably as a prelude to a large scale artillery barrage.

(2) The destruction of the balloons was obviously to cut off American observation, probably of German movement in the rear.

(3) Items (3) and (5) indicate careful reconnaissance both by observation and aerial photograph. The fact that regular patrol reconnaissance was carried on in the same area indicated preparations for movement on the part of the enemy.

(4) Both sides knew of the gap in the wire, and under normal conditions, the gap would have been closed. In this case, it was left open, but camouflaged, indicating a group movement through the wire or a trap set by the enemy.

(5) The shrapnel in ranging bursts was undoubtedly in anticipation of closing off the road to the American rear, and to prevent bringing up the reserves.

These five deductions resulted in a decision by the commander to strengthen the sector to oppose an enemy attack. The anticipated attack did materialize and was repulsed with heavy German losses.

Accurate information is vital. The trained observer will report only what he sees. His personal opinion on what he saw, or his own interpretation of what he saw must not be confused with the actual facts. Many untrained observers report their deductions rather than their observations, a practice which must be stopped at its inception during the training period.

An incident as the Germans swept through the lowlands in 1940 illustrates this point. The Belgians held one bank of the Scheldt River, where they set up OP's to observe any German action. Suddenly an oil tank on the German side went up in flames. The OP personnel thought it was sabotage, to reduce the enemy's limited supply of gasoline. Therefore, they deduced that it was favorable to the Belgians, and *they did not report it.* Their interpretation of the occurrence proved disastrous. Under cover of oily smoke that settled on the water German motor boats carrying troops crossed to the Belgian side. Here deductions were fallacious. Actual observation, not interpretation must be reported back by the observer.

Too positive identification of objects or incidents can also be a dangerous habit, especially when the observer is lacking in experience. Many men have a tendency to see an object, a piece of artillery, for instance, and positively identify it as a gun of a certain size and caliber when in reality they do not know what size it is and are just guessing.

In some cases such positive identification may have to be taken at face value at the command post. They may lead to dispositions of troops and changes in tactics to meet enemy threats based on these positive identifications. There is a great deal of difference between a 37mm AT gun and a 105 mm Howitzer, yet many WW II observers positively identified 37mm AT guns as 105m Howitzers and vice versa. Whenever uncertainties exist concerning a positive identification of objects, the words such as "estimated to be", "appears to be", etc., should be used in reporting it. For example:

"Enemy arty. piece *estimated to be* 105 Howitzer seen at RJ 206."

Figure 33

WW II battle experience in Tunisia showed the inability of many untrained observers to report accurately what they saw and heard from an OP. Trailers towed by large trucks were reported as artillery pieces, haystacks were mistaken for tanks, and groups of Arabs tending flocks of goats were called German patrols.

On one occasion a vivid imagination misconstrued a small fire or candle burning in an Arab hut as an enemy column moving west. Checking this, it was found to be several lights, which were mistaken for blackout lights on enemy vehicles. Such errors in observation can only be corrected by training and will not be uncommon with untrained personnel.

Special observation situations such as operations in the jungle, snow, or high mountains will necessitate some changes in technique and equipment, but the basic fundamentals covered here will apply in almost all types of observation conditions.

Selection of the OP Site

The proper selection of the observation post is of primary importance. It will ordinarily be situated on commanding ground which affords the best view of the terrain towards the enemy and in the rear of his front lines. In open warfare, where the lines are not fixed, the OP will ordinarily be one of temporary nature. The scout or patrol member will select a point on the terrain where cover and concealment is provided by natural cover or terrain irregularities.

In a stabilized situation the OP will be more carefully selected, camouflaged and constructed for use during that period of the sector's importance. Such an OP will be constructed to provide all possible cover and concealment from enemy artillery, small arms fire and bombing. Ideally it will be large enough so that men can be comfortable while observing and it will probably contain more elaborate equipment such as range finders, periscopes, radios, phones, etcetera, than will be used in a temporary patrol type OP.

An observation post site should have good visibility, a covered route of approach and be camouflaged and concealed, both from the ground and air. Elaborate camouflage is usually reserved for permanent observation posts, but no observation post, whether it is to be used for ten minutes or for ten days, can expect to remain in operation very long if visible or obvious to enemy eyes. It should also be recognized that location of the OP by the enemy does not necessarily mean that he will wipe it out at once. He may wait until an action starts before destroying it, thus blinding the unit when it needs it's OP most.

Concealment from enemy ground observers is not enough, since it is easy to locate an observation post, gun position, or command post from the air by following tracks or a beaten path. Such clues stand out on an aerial photograph like arrows pointing to the hidden post. Photographs will reveal OP positions by showing loose dirt, wire lines, paths, and approaches which end at the

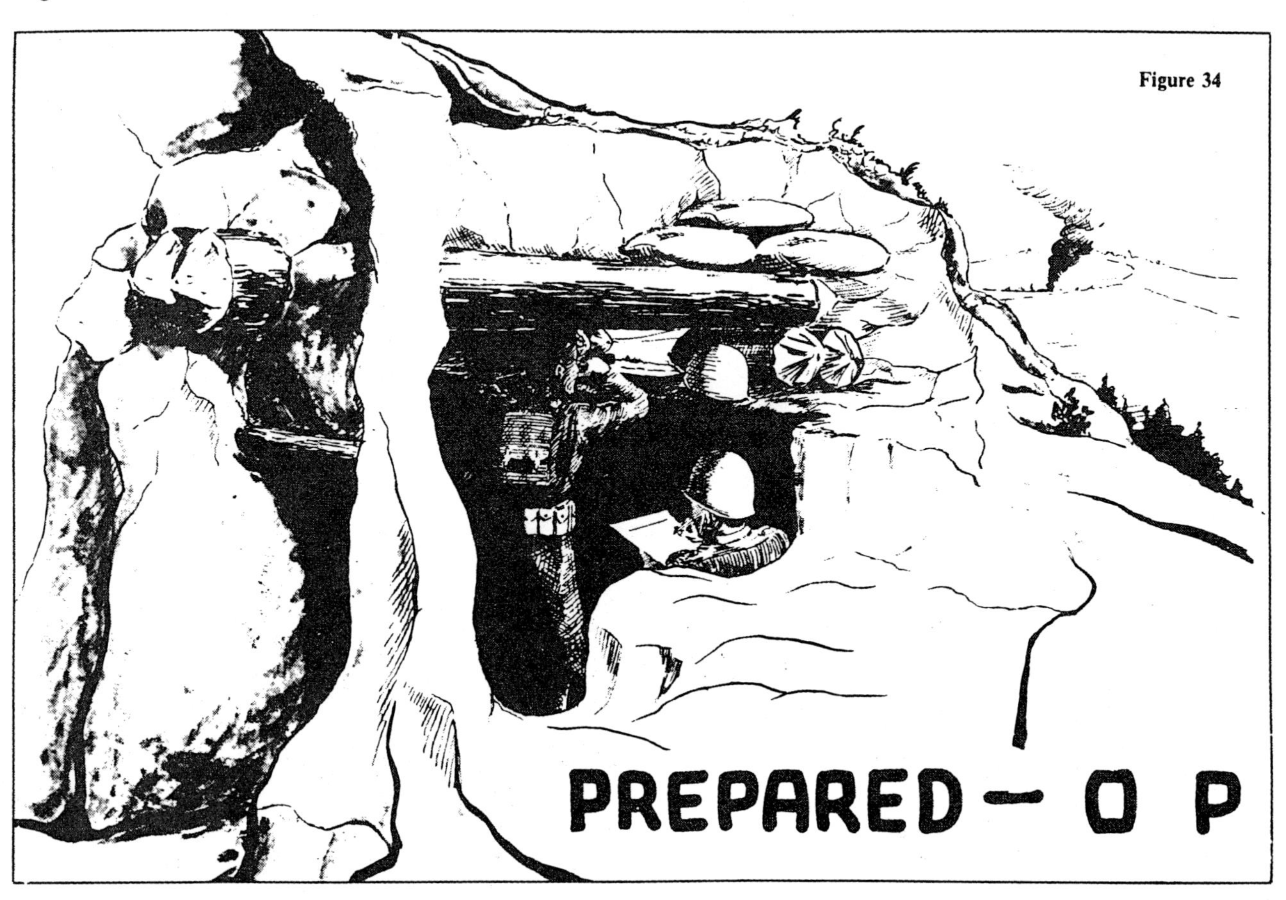

Figure 34

Figure 35

post. An approach independent of an existing route (path or trench) must be prolonged past the post or be camouflaged. There must be complete camouflage discipline. Any breach of this discipline may lead to discovery.

Full consideration in choosing the location for an observation post must be given to concealment from the eyes of the enemy who will look at the OP from the *front.* Many OP's must be chosen at night for daylight observation. On such occasions particular attention must be paid to concealment of the OP. Just because it is dark and little personal concealment is necessary does not mean it will escape detection by the enemy in daylight.

Our forces in Tunisia learned early to dig OP's at night, to make reliefs under cover of darkness and to hold movement to a minimum throughout the day. Otherwise German artillery soon picked them up, and it was necessary either to establish another OP or to do observation between concentrations.

The Germans and the Japanese always tried to deny observation to their enemies. Within individual sectors or on open flanks they searched by fire or combat patrols those positions most logical for opposing OP's. Reports from the European theater stated that the Germans always searched skylines and military crests for OP positions. Peaks and high points beyond the scope of combat patrols were searched by machine gun and mortar fire. In Tunisia, to establish OP's either on the crest of the hill or on its most prominent peak was found to be suicidal. Enemy counter-measures were successful if faulty discipline in concealment, camouflage, tracks, refuse, or just plain carelessness betrayed the presence of the OP. The enemy must be outsmarted.

The specific location of an OP should be in an area the enemy believes is unsuited for observation. OP's should avoid distinguishing landmarks, such as lone trees, a farmhouse, a sharp crest, towers, summits, paths other than the side paths, or a place just below the military crest of the hill. The best positions are inconspicuous, for example: in a forest or woods, in the midst of an irregular hedge, row, or in broken, irregular ground. There should be a covered route of approach and withdrawal to make the good observation post site complete.

Whenever OP's are sited in exposed areas they ordinarily will not remain in operation long and their personnel may become casualties. Because of this a second rate OP, placed where it will function during any type of friendly or enemy action, may often be better than one which will be overrun or wiped out in the first phases of battle. Care should be taken that there is liaison or communication established with adjoining OP's especially if part of the sector is marked by a hill or ravine. The OP should be close as practical to the C.P., but it must be far enough away so that fire directed at one will not cover both.

Generally the OP will be located either on the forward or reverse slope of a hill. Both have their advantages and disadvantages.

Reverse Slope: OP Site Advantages

(1) Greater freedom of movement is permitted during the day.

(2) It is easier to install, maintain and conceal communication facilities and installations.

(3) It can be initially occupied and sited during daylight.

(4) It is possible to get greater depth observation into the hostile area.

Figure 36

Disadvantages

(1) Enemy fire adjusted on the crest of the hill endangers the installation.

(2) OP personnel, radio antennaes, etc., are difficult to conceal as they must come to the crest to observe. This disadvantage is most apparent when the crest of the hill is sharp and objects are clearly silhouetted against the sky. It is minimized when the hill used blends into a larger one in the rear, causing the skyline to be less distinct.

Forward Slope: OP Site Advantages

(1) Concealment is easier if a background is present, preventing silhouettes.

(2) A better view of the hostile area to the immediate front is possible.

(3) A forward OP may be far enough down from the crest that enemy fire directed on the crest will not touch it. To neutralize such an OP the enemy will be forced to cover the entire forward slope with fire if he can not pin-point the location.

Disadvantages

(1) Signal communication in daylight is difficult.

(2) Location change and movement is not possible in daylight.

(3) Must be occupied under the cover of darkness to prevent discovery. The occupation and selection of an OP site under the cover of darkness is difficult and a careful terrain study must be made to insure that it will be usable in daylight.

Whenever possible, the mission of observation should never be entrusted to a single OP. At least two OP's should be established for the following reasons:

(1) To insure all around continuous coverage.

(2) To check and confirm one another's work. (The men in the OP's will then be able to locate accurately the position of weapons by sound and flash or by intersection.)

(3) If one is destroyed or neutralized, the other will still maintain observation.

Alternate OP's must be located and prepared at the time of establishment of the principal one. Possibility of neutralization by smoke, fire, raids, or by weather must be considered in the selection of the alternate site. The alternate OP should, like the main one, be so located that continuous observation will be insured regardless of the action. A covered route from the principal OP to the alternate must be considered when the site is selected. Hasty, poorly concealed evacuation due to enemy action will disclose the location of the alternate site to him.

If certain areas in an assigned section are defiladed from the chosen OP site, auxiliary OP's must be set up to cover these areas or observers in other sectors must be given the responsibility of covering those areas hidden from view.

Equipment

Basic OP equipment will include: field glasses, compasses, watch, maps, or aerial photos, material for recording observation such as special OP report forms. overlay paper, message books, pencils, and a means of communication.

The OP is useless unless a swift means of communication is available. The field phone is the best all around means of communication. Often wire will be laid concurrently with the selection of the site. When wire is laid, it is desirable to lay two separate lines into the OP so that enemy action will not delay the transmission of information. Consideration must also be given to the laying of wire to the alternate site so that is can be in instant use when conditions force its occupation.

Flares, pigeons, messengers, radios, panels (reverse slope), and colored smoke grenades (rifle) furnish additional and emergency means of communication with the parent unit. Sound signals of various types, dogs, tracers, semaphore, and signal lamps may be used under certain conditions.

OP personnel must be able to defend themselves against enemy action. Small arms and grenades must be ready for any emergency. Food, water, and entrenching tools must also be considered in preparing and operating an OP.

Organization

Organization of a stabilized OP is progressive in nature, Immediate observation is established while the exact point to be occupied is selected. As soon as observation is established, the definite locations of the OP and the alternate are relayed to the parent unit by the swiftest and surest means of communications.

After initial occupation and communications are completed, the observer establishes a base orientation line to some easily identified object in his assigned sector. A range card is made up showing aximuths and distances to important points on the terrain. All photos and maps are marked accordingly. Once this has been done the observer can quickly locate and report any action in his sector.

A panoramic photo can be employed to facilitate locating enemy activity even more accurately. These photos cover the terrain to the front of the OP and are set up in the OP's with ranges and azimuths marked on them. A variation is known as the "gridded oblique" which is an oblique line overlap taken at 1,000 to 4,000 ft and normally from behind our own lines. Instead of actually marking the photo, a special grid is superimposed to provide a means of referring to points, and of finding their map positions and heights.

The Germans used panoramic photos in their OP positions during the last war, and continue to do so now. Their importance and efficiency is very great. Accuracy as well as speed in reporting are greatly enhanced.

S-2's and unit commanders must recognize that the operation of an observation post is a dull and tedious job where it is very easy for a man to become careless, inattentive or to fall asleep, especially if he is battle weary or physically exhausted. Shifts must be changed frequently and observers must be given the chance to rest if good accurate work is to be expected.

Men operating the observation post usually work in pairs, one to observe the other to record the information obtained. These two should alternate every 15 minutes to half hour and should always be close enough together to permit conversation in low tones. It is best not to observe continuously for more than one hour at a time. Longer periods cause the small details of change, which are so important, to be over-looked. Ideally the pair should be relieved every few hours, but if only two men are available for this work, they must work in shifts - one observing and recording, the other sleeping.

A ground observer's report sheet can be improvised to be used by the recorder to set down the time, place, character, etcetera, of all observations. Recurrences or discontinuances of events recorded on this sheet may be of prime importance. Events of sufficient urgency are sent by the swiftest means to the C.P. Accumulated observers reports are handed in to the S-2 at stated intervals.

Whenever possible, the same observer should be used repeatedly over the same area. He will learn the details of his particular area so thoroughly that he will instinctively notice any change. Therefore, the best plan is to assign the same men to a definite observation sector for the duration of that sector's importance.

The observer locates himself in a position which affords steady rest for his glasses and from which he can secure the best view of his sector. He then mentally divides the area included in his sector of observation into a series of overlapping zones, the nearest of which takes in the ground just beyond the front line of his unit, while the farthest includes the limit of practical visibility. Beginning with the zone nearest them, the observer makes a slow and thorough examination of the terrain, searching from one edge of the zone to the other. Proceeding to the examination of the next zone, he searches it in the opposite direction, and he continues in this manner until the whole sector has been examined. He searches for indications of the enemy such as trenches, paths, gun positions, OP's, wire, troop movements, etc. If any movement or unusual feature catches his eye, the point where it occurred should be watched closely at regular intervals.

Figure 37

When a definite, well bounded sector has been assigned for the observer by his S-2 he becomes responsible for everything in that sector. If enemy action is observed outside his zone of responsibility he reports it, but it should be considered secondary to any activity in his own sector. If two or more OP's are used per unit, the sector for each OP slightly overlaps that of the adjacent OP. Observed incidents and enemy activity should be recorded in a locally improvised ground observers report form. These forms should be turned in when shifts are changed.

It is good practice to establish regular routine and time intervals to report ordinary activity to the parent unit. Negative information should be reported at these times as well as positive. Enemy action of sufficient urgency will be transmitted at the time it is observed.

In relieving OP personnel a "time overlap" is necessary so that outgoing personnel can thoroughly familiarize the newcomers with the area, the section that is

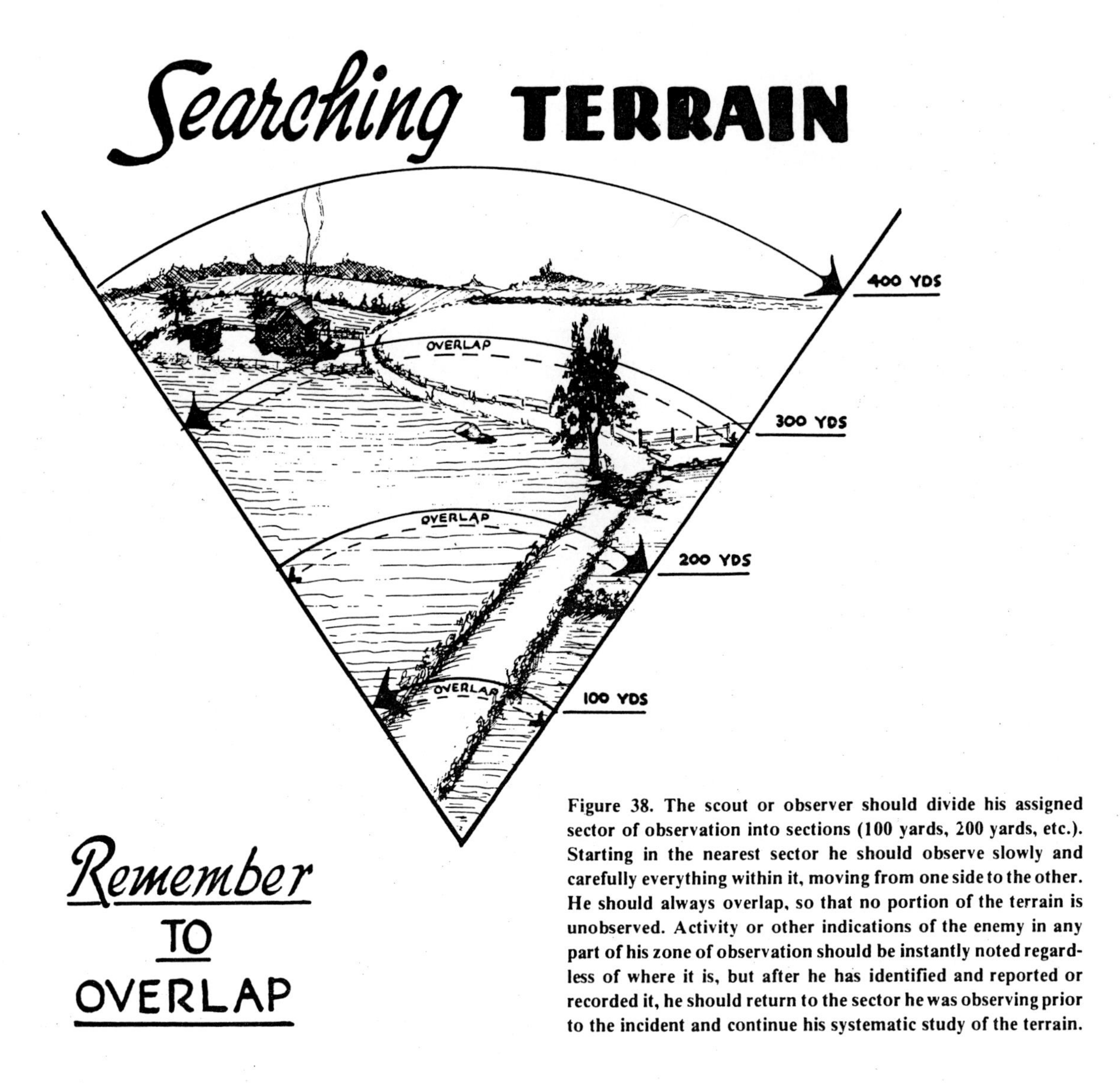

Figure 38. The scout or observer should divide his assigned sector of observation into sections (100 yards, 200 yards, etc.). Starting in the nearest sector he should observe slowly and carefully everything within it, moving from one side to the other. He should always overlap, so that no portion of the terrain is unobserved. Activity or other indications of the enemy in any part of his zone of observation should be instantly noted regardless of where it is, but after he has identified and reported or recorded it, he should return to the sector he was observing prior to the incident and continue his systematic study of the terrain.

going on, previous action, and enemy concentrations. New observers should go out with an experienced observer before they take over a sector on their own.

Security

The OP should have security guards where necessary. There is, of course, a distinction between the patrol OP, which must furnish its own security, and the field or unit OP, which should have personnel assigned for specific duty as security.

In night operations it is usually sufficient if one man is awake at all times. Generally, OP personnel should fall back at night unless they are extremely well concealed or are adequately protected by the infantry. The enemy is usually most active at twilight or dawn and observers should be especially careful at these times. The enemy will often dispatch special patrols to locate and destroy observations posts.

Besides the commanding officer and his staff who will frequently make use of an OP for tactical purposes, there are other commanders, higher commanders, observers, visitors and newspaper men who may frequent the OP. If these visitors are unavoidable, and can be expected, the OP should be dug or otherwise made large enough to accomodate them. Security of an OP is often violated by misuse due to over crowding or failure of the visitors to observe covered routes of approach and departure. Such conditions will occur particularly when observation is limited and only one OP has observation on a sector. The effectiveness of an OP can be destroyed by conditions of this type. Ideally, the OP should be restricted to operational personnel and local commanders only. A special visitors OP should be constructed if conditions warrant.

What to See

Training in observation is not complete without instruction in "what to see and what to report." Each observer must learn that it is his job to report *everything* he sees or hears. And he must be trained not only in what he should expect to see but also what he should expect to hear. The observer must realize that every bit of enemy information obtained will contribute to the success of his own troops. It is criminal for him to risk his own life in the successful accomplishment of a mission and then fail to bring or report back information of something observed simply because his mind had not been trained to register it or because he did not know what to look for.

THE OBSERVER SHOULD MAKE OUT A RANGE CARD AS SOON AS POSSIBLE AFTER THE 'OP' IS SET UP ~ THE REFERENCE POINTS, KNOWN RANGES, AND AZIMUTHS ESTABLISHED ON PROMINENT POINTS WITHIN THE ZONE OF 'OP' RESPONSIBILITY FACILITATE THE RAPID AND ACCURATE LOCATION AND REPORTING OF ENEMY INSTALLATIONS AND ACTIVITIES.

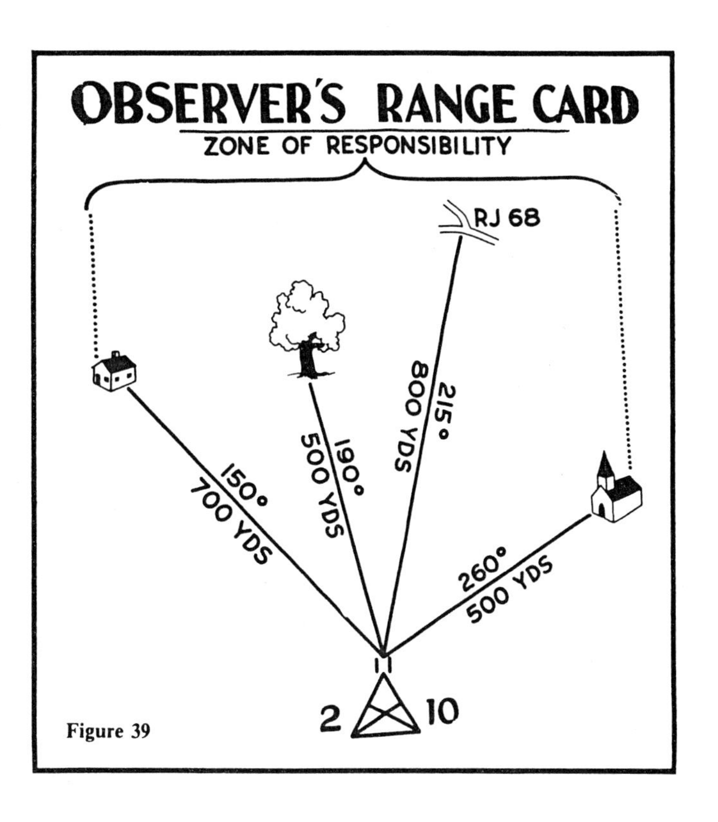

Figure 39

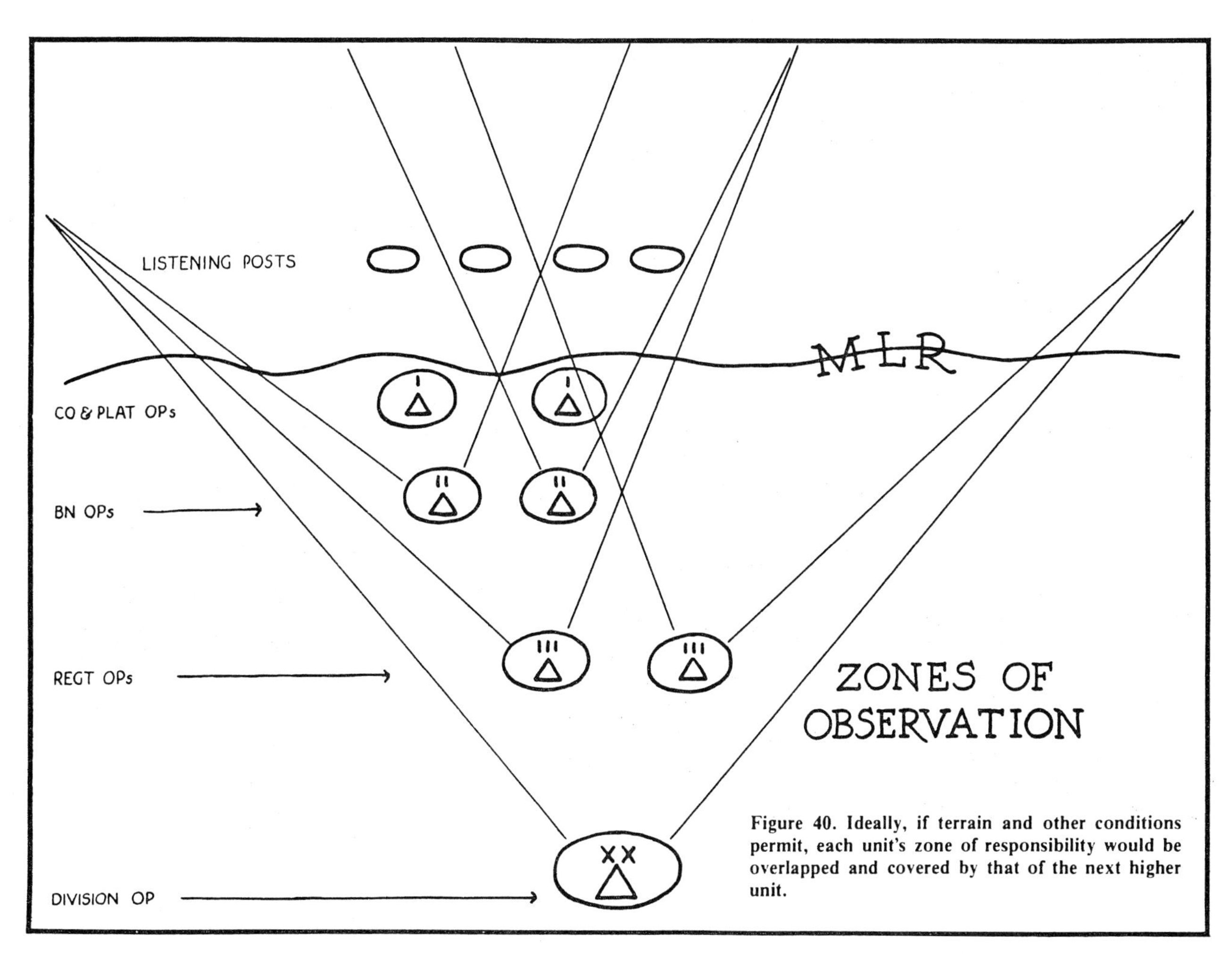

Figure 40. Ideally, if terrain and other conditions permit, each unit's zone of responsibility would be overlapped and covered by that of the next higher unit.

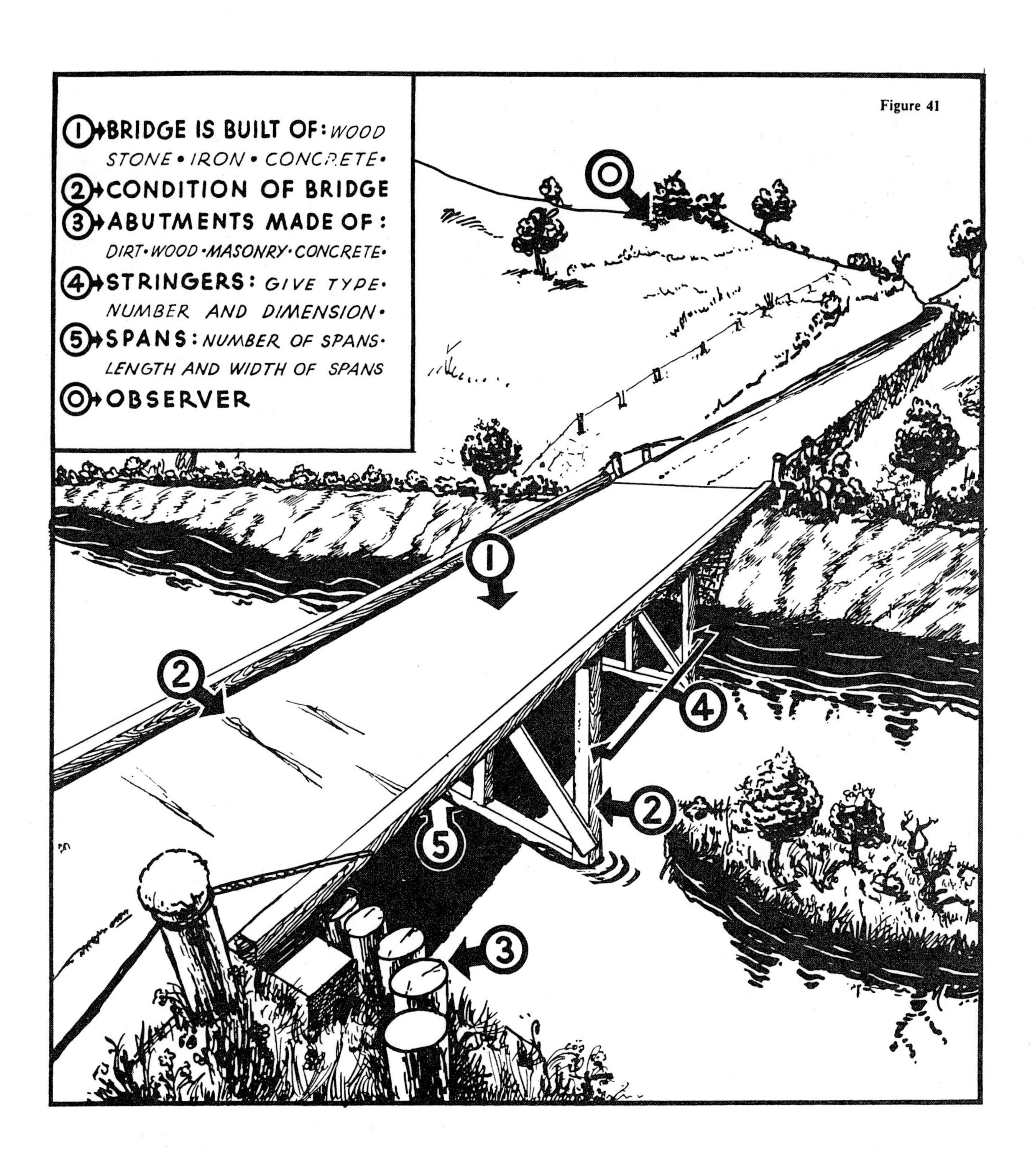
Figure 41
①→BRIDGE IS BUILT OF: WOOD
STONE • IRON • CONCRETE •
②→CONDITION OF BRIDGE
③→ABUTMENTS MADE OF:
DIRT • WOOD • MASONRY • CONCRETE •
④→STRINGERS: GIVE TYPE •
NUMBER AND DIMENSION •
⑤→SPANS: NUMBER OF SPANS •
LENGTH AND WIDTH OF SPANS
Ⓞ→OBSERVER
1
2
2
3
4
5
O

The observer must learn through training what he is to look for and must be able to interpret everything coming within range of his sight, hearing and smell that pertains to the enemy. Both the observer and the scout must be able to read the natural signs on roads, streams, and terrain to determine enemy activities. They must learn to estimate enemy strength as well as dispositions. To do this they must know the size of enemy units, their length in column, and the time it takes to pass a given point. They must have practice in this type of estimation. Their messages must be concise and clear, their overlays accurate.

To be able to interpret enemy signs, activity and dispositions the observer or scout must have a complete knowledge of comparable activity, dispositions etcetera, of his own forces. He must first be able to recognize his own army's equipment, (trucks, tents, vehicles) armament, (small arms, artillery, and troop units). After he has been trained and tested in this type of recognition he will be better able to recognize that of the enemy, as he will more readily note similarities and differences. Such training must be given him during the initial phase of his instruction.

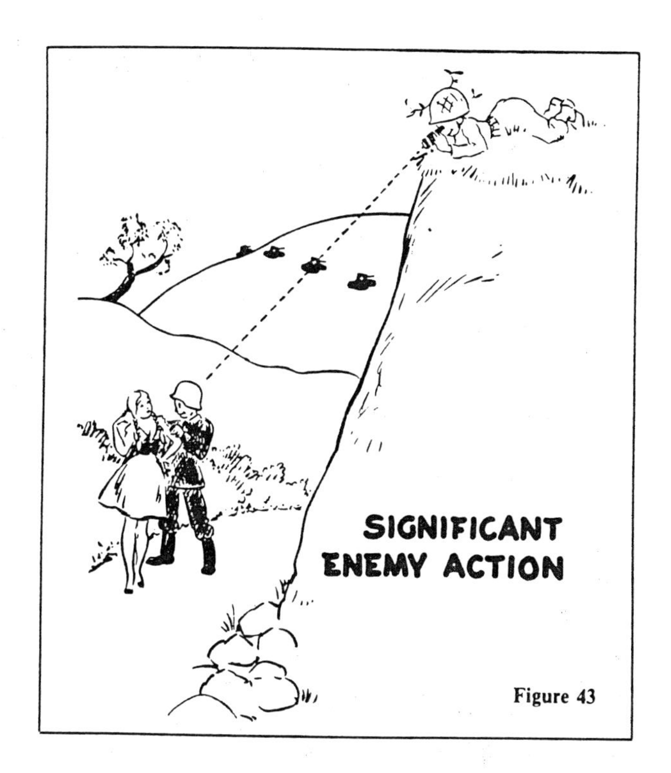

Figure 43

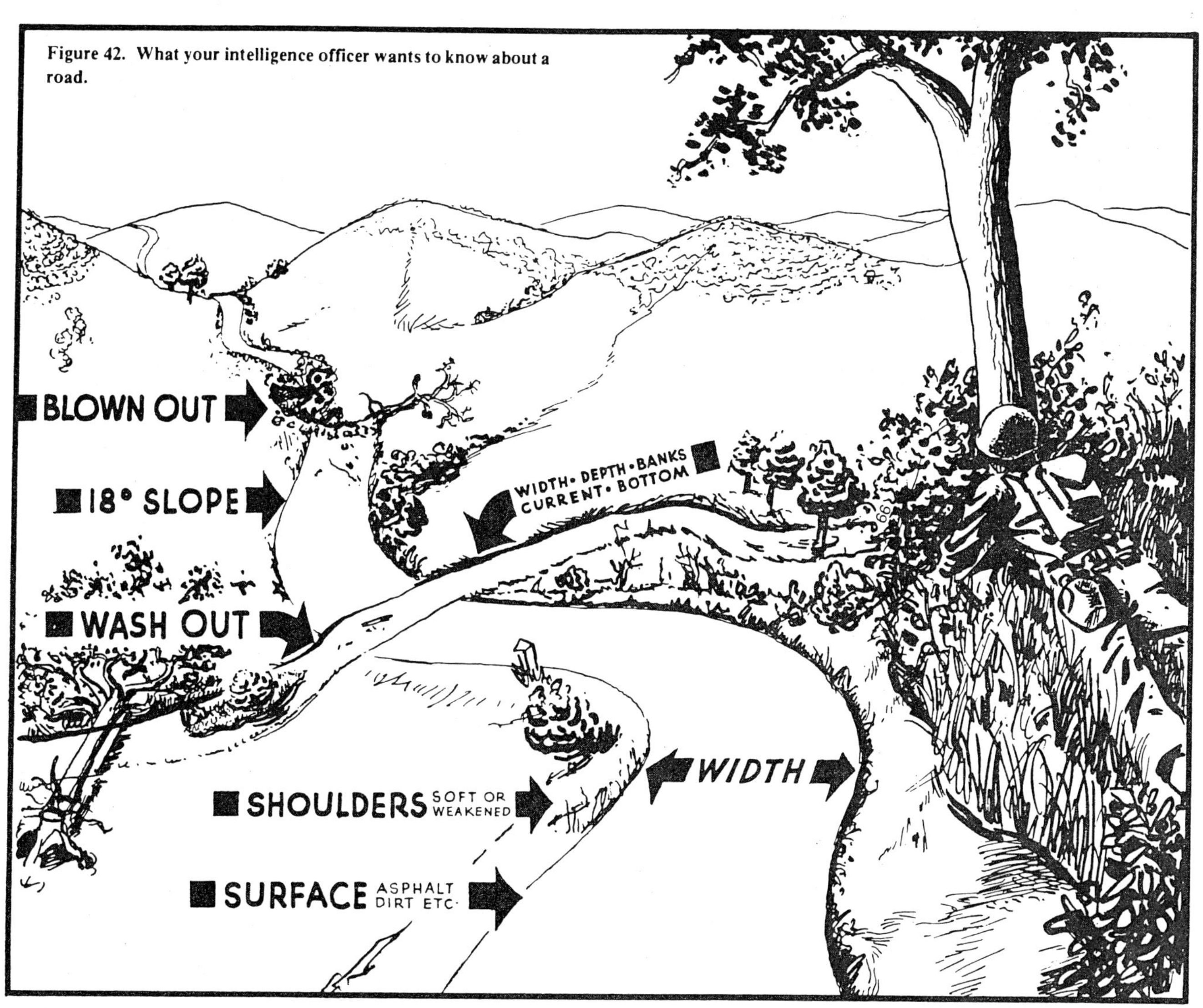

Figure 42. What your intelligence officer wants to know about a road.

Figure 44

Untrained observers and scouts show little conception of what is asked of them. They report things that are easy to note and frequently neglect to note matters of importance required for operations. For instance, enemy troops may be concentrated on the opposite side of a particular bridge. The presence of this concentration will be duly reported, but no report will be given on the following factors:

(1) *About the bridge:* how low and how wide? Was it mined? What kind of traffic could use it?
(2) About the *terrain* before and beyond the bridge? What kind of cover was on the approaches to the bridge?
(3) *About the river:* How wide and deep is it? Are the banks steep or can they be climbed? How far is this spot from the enemy?
(4) *About the hills:* What hills are along the way? How high?

A Check List for the Observer

(1) Can I go to the observation post at night or in fog and rain? Do I have proper maps and photos?
(2) Am I keeping the zone of observation under constant surveillance?
(3) Have I selected and prepared an alternate observation post?
(4) Have I made a sketch of my zone of observation and a range card?
(5) Is my camouflage perfect or can I improve it?
(6) Do I have plenty of rations and water, arms, and ammunition?
(7) Is the security sufficient?
(8) Are communications in good working order? Do I have a supplementary means of communication?
(9) Have I reported everything as observed?

LISTENING POSTS

An observer can see very little at night and therefore must rely chiefly on his hearing to obtain information. If it is necessary to move about, the observer should stop frequently and listen intently. The most practical means of night observation is through the establishment of a listening post. Such positioned out guards and listening posts can be replaced by small patrols in some sectors.

The front of any defensive or static position is usually covered by listening posts. They are normally located in open ground or in pits, which are camouflaged during the day. Under cover of darkness, the observer

Figure 45

Figure 46

crawls as close to the enemy lines as possible. In general, listening posts are closer to the enemy than observation posts and where practicable, are located on low ground. The listener lies prone with his head to the ground. He is thus best concealed and occasionally may be able to see objects silhouetted against the sky. The scouts or observer should be able to interpret sounds heard at night such as troop movements, tanks, movement of patrols, digging, and location of artillery fire. The listener then determines the aximuths, estimates ranges, records or reports his observation by radio or phone.

The listening post must be operated by courageous, skilled men as they are always subject to counter measures by enemy patrols. The two-man type is most often used for security and psychological reasons.

The results of listening post observations are checked and confirmed by daytime observation from ground and air.

ARTILLERY OP'S

Artillery observation posts may be near or may be located with the infantry OP's. This is dependent on local terrain conditions and the requirements of observation. The artillery observer must always be considered an additional source of information about the enemy and must be trained to report accurately all information even though not used for artillery purposes. Infantry observers at times may have to direct artillery fire on enemy targets.

The adjustment of artillery fire by an infantry observer can be accomplished by allowing the artillery fire location center to make all computations and send all fire commands to the batteries. This procedure is one frequently used because the artillery forward observer can not cover all sectors at once, and the possibility is always present that he and his team may become casualties. If artillery is supporting an infantry unit an artillery liaison officer can usually be contacted at the CP (Regiment and above). This officer can be contacted to establish the necessary connection with the firing battery. The position of the OP can then be given. The nature and location of the target may be given by map coordinates or by range and azimuth (polar coordinates) from the OP site. After the bursts have been corrected by giving the amount of yards shot over or short and right or left and the fire has been placed on the target, the command "fire for effect" can be given.

RANGE ESTIMATION

In addition to being able to see and recognize an object, the scout or observer must be able to locate it by using his compass and by accurately determining the range.

There are three basic methods of range estimation, eye, sound, and mechanical. All three methods should be covered in training.

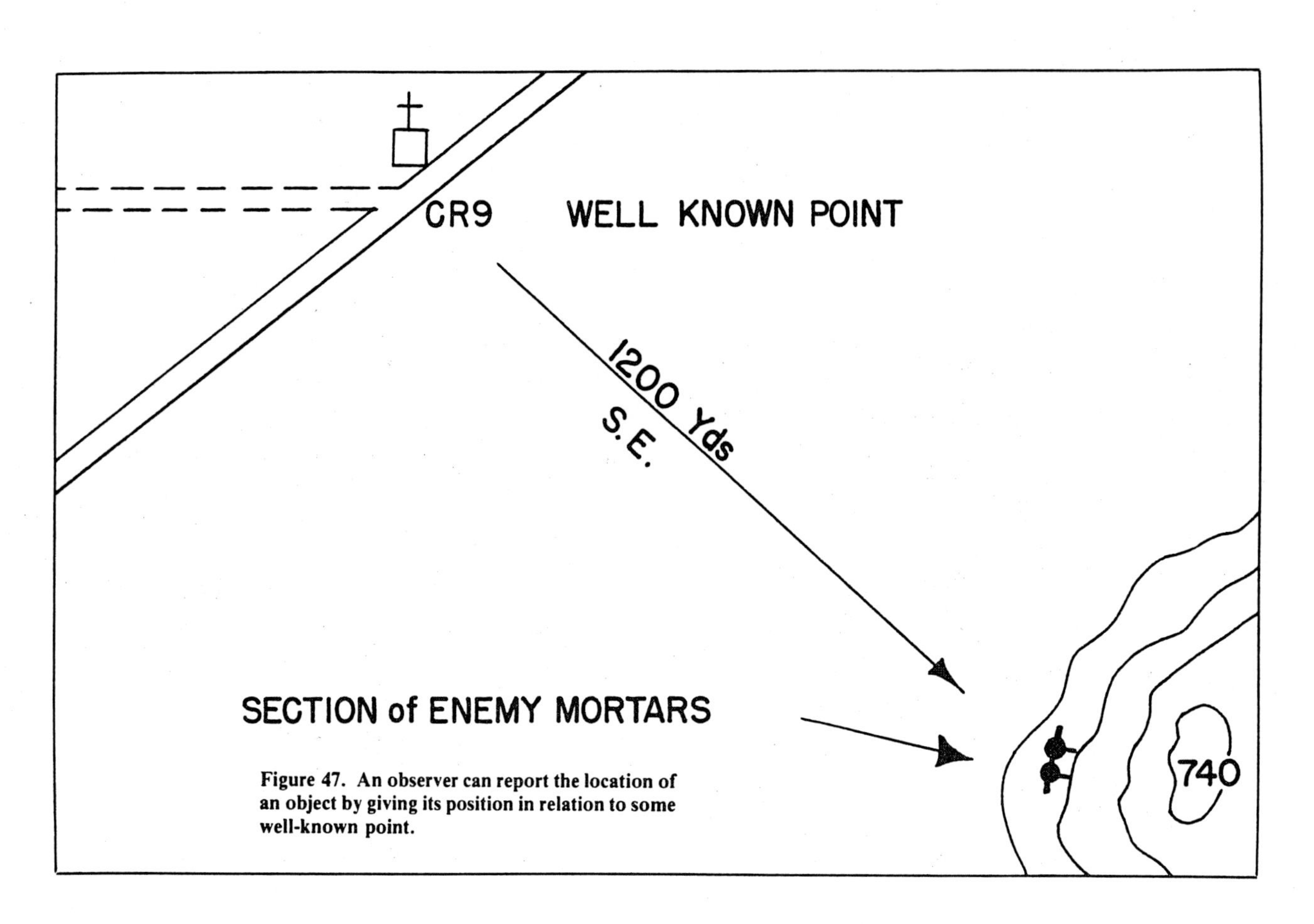

Figure 47. An observer can report the location of an object by giving its position in relation to some well-known point.

Figure 48. The scout should know simple time and space tables so he can give a good estimate of enemy strength.

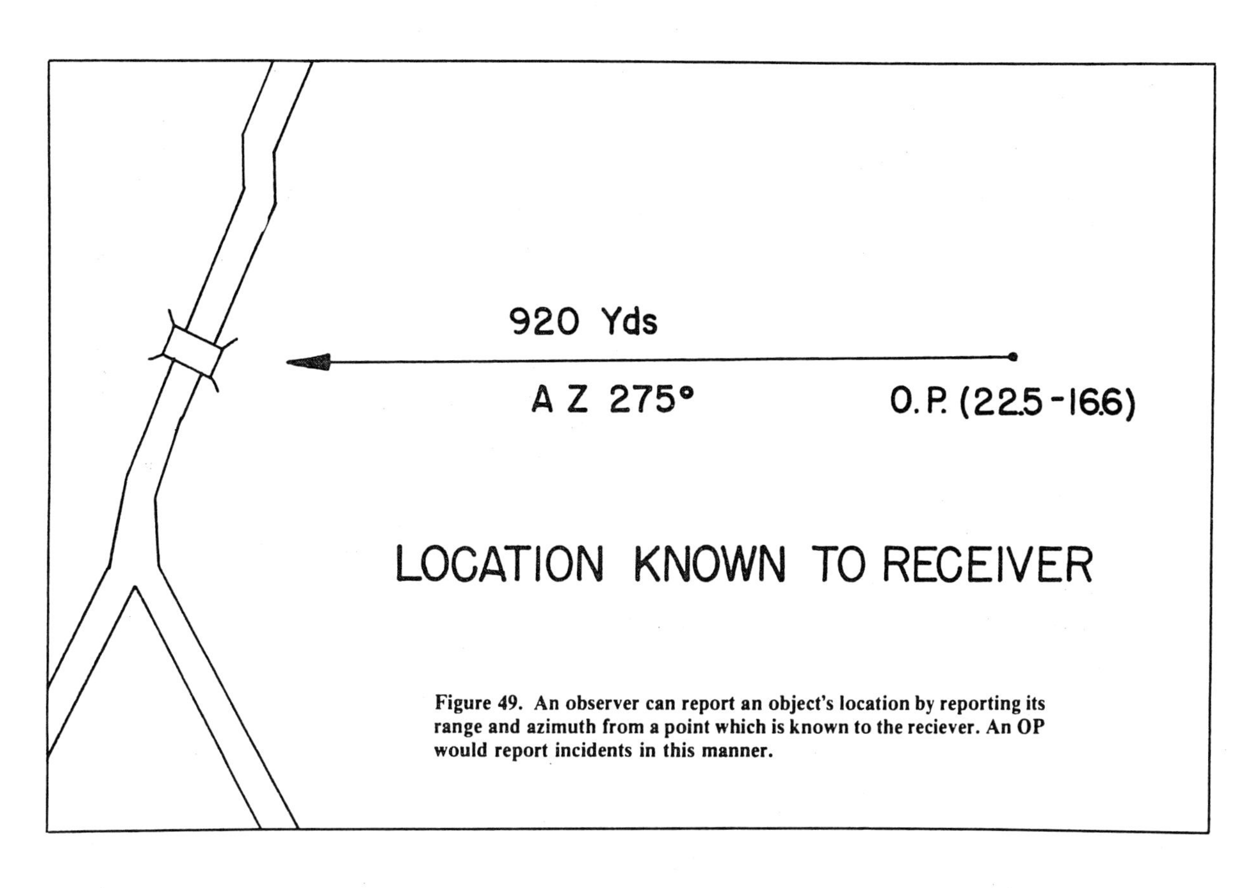

Figure 49. An observer can report an object's location by reporting its range and azimuth from a point which is known to the reciever. An OP would report incidents in this manner.

Figure 50. This illustration shows how to locate incidents at night by means of sound and flash.

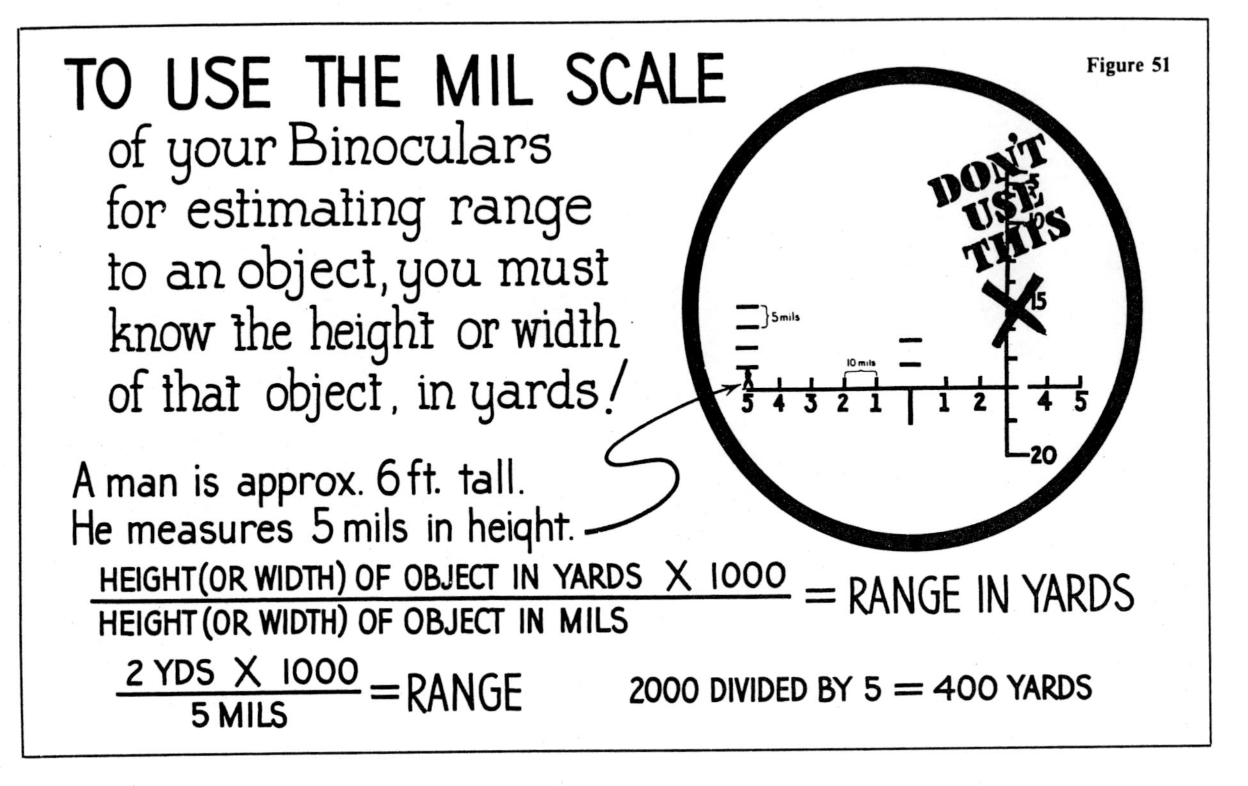

Eye

Range estimation by eye is a skill that can be acquired only with constant practice. Accurate range estimation by eye is especially difficult when terrain is irregular and distances exceed 500 yards. The application of the "mental unit of measure" method is helpful in the early stages. If the observer has in his mind a mental picture of the length of some familiar object (a football field for instance) he can take the distance of this mental picture and place it between himself and the target. By using this mental unit of measure like he would a yard stick he can estimate up to 500 years. When the range is obviously between 500 and 1000 yards, a point half-way to the target is selected. The range of the half-way point is determined by applying the mental unit of measure and the result is multiplied by two to give the desired range.

When much of the ground between the target and the observer is hidden from view, the mental unit of measure cannot be applied. In such cases, the range is estimated by "appearance of objects". Conditions of light and terrain have considerable effect upon the appearance of objects, making them appear nearer at times and at times more distant then they really are.

How Objects Appear

Whenever the appearance of objects is used as a basis for range estimates, the observer must make allowance for the effects noted below:

(a) Objects seem nearer:
1. When the object is in bright light.
2. When the color of the object contrasts sharply with the color of the background.
3. When looking over water, snow or a uniform surface, like a wheatfield. If the individual can estimate range on land accurately he can do the same on water by by using the following method:
Place the back toward the water and bend forward and look at the object through the legs. Because of the apparent changed position, the water and sky estimation of the water surface is the same as like distance on land.
4. When looking down from a height.
5. In the clear atmosphere of high altitudes.
6. When looking over a depression, most of which is hidden.

(b) Objects seem more distant:
1. When there is little light or a fog.
2. When only a small part of the object is seen.
3. When looking upward from low ground toward higher ground.
4. When looking over a depression, most of which is visible.

The average of a number of estimates by different men is more accurate than an estimate by an individual. Hence, in firing exercises and in combat, if conditions permit, the average of the estimates of several men is best.

Sound

To estimate distance by sound, the "crack and thump" method may be used. This is based on a knowledge of the velocity of sound which is 1080 feet per second (360 yards).

If an artillery shell is observed to hit in an area and two seconds elapse between the dust or flash of impact and the sound of the detonation, the range is 2 x 360 = 720 yards.

Seconds x 360 = Range in yards.

The muzzle flash (or smoke) of an artillery piece observed before the sound of the explosion (propelling charge) is heard can also be used for sound estimation.

Mechanical

Range finders such as those used by artillery forward observers or those captured from the enemy can be employed if available. Unfortunately, these excellent instruments are only available in limited quantity. However, the scout or observer can utilize the mil scale on his binoculars for accurate range determination.

The basis of this means of estimation is a knowledge of the length, width or height of common objects, such as houses, automobiles, roads, railroads, etc. Knowing the size of these objects, the student then applies the simple formula of the mils scale as applied to linear measurements.

Applications of this principle are simple. All military field glasses have crossed hair lines on which is imprinted the mil scale. On the object of known size, place the mil scale of the hair lines and read the number of mils. Assume for instance, a house ten yards high (since American military distances are measured in yards, all calculation should be computed in yards) covers the distance of four mils on the scale. What is the distance to the house?

$$\text{RANGE} = \frac{\text{Known width of house (yds) x 1000}}{\text{Width in Mils (Glass Measurement)}}$$

$$\text{RANGE} = \frac{10 \times 1000}{\text{4 Mils (Glass Measurement)}} \qquad \frac{10000}{4} = 2500 \text{ yards}$$

MESSAGES

Modern means of communication have greatly simplified the problem of communication between the scout patrol or observer and his unit. Light-weight radios, sound powered phones, flares, pigeons, signal lamps, are all available for communication. These methods of communication do not eliminate the need for the written message, but in many cases they provide a surer, swifter means of transmission of information from reconnaissance to their unit.

However, these modern aids in the transmission of information will never entirely eliminate the necessity for the written message. Mechanical failure, local enemy situation, security, transportation and supply difficulties make it necessary that all reconnaissance personnel be able to write clear, complete concise messages for delivery by messenger, when no other method of transmission is available.

Proper message writing from the staff viewpoint can and does become a very complicated process, but from the standpoint of the scout a good field message needs only the 5 W's (who, what, when, where and why.) Special message forms are not always available but any scrap of paper containing the 5 W's will be complete.

(Here is a simple guide or check list for all field messages):

WHO — is writing the message? What patrol number? Identity of individual or unit, patrol, etc.
WHAT — was the enemy doing? If moving, in what direction, about what speed?
WHEN — did you see it: The time when you saw the action unless it coincides exactly with the time the message is signed.
WHERE — did you see it? The exact location - given by reference points, coordinates, overlay, etc.
WHY — What is the basis for your statement? Why do you report this?

CHAPTER IV

PATROLLING

"The patrol is a hunt in which the hunter can become the hunted."

Patrolling is a direct reflection of an army's ability or inclination to fight and has been called the art of "infiltration" by the infantry. The function of patrolling in its relationship to modern battle has been likened to the left hand of a boxer, the patrol probing the opponents defenses striving to find weaknesses and screening intentions until the right hand or the main body can deliver the knockout.

It is a long established precept, recently re-emphasized, that an army on the offensive functions in direct proportion to its patrolling skill. In the fluid and jungle type of warfare, which has been characteristic in many campaigns of recent date, patrolling actions have been responsible for a major part of the actual operations. In static situations the aggressive spirit and control of No Man's Land have only been maintained by active patrolling.

The patrolling concept of 1918 was to consider it generally in the light of close-contact operations found in trench warfare. WW II campaigns in Africa, the Pacific, Russia and our involvement in Vietnam have served to stress longer range phases of patrolling as they affect actual combat and operational planning.

The theory of ground reconnaissance, which emanated from the First World War advocated the use of highly skilled individual scouts or small patrols to penetrate the enemy lines and get information. They fought only in self-defense and operated by stealth. Experience has shown that this type of reconnaissance alone is not sufficient to meet all the demands of modern war. Generally, present day patrols can be classed as one of two types: a patrol small enough to sneak, or a patrol large enough and strong enough to fight. Individual scouts and small, lightly armed, reconnaissance patrols always have their uses but they also have limitations. The sneak patrol can only gain information by observing and listening. When the enemy is silent and well concealed, such a patrol becomes an ineffective means of ground reconnaissance. The use of large, well-armed patrols carrying a great amount of automatic fire power, for reconnaissance as well as for combat, is now standard procedure in all armies. They fulfill a need that the small sneak patrol is not capable of meeting. Although this type of patrol sacrifices some of the advantages of concealment, stealth, and silent movement which are inherent in the individual scout or small patrol, it is better able to operate in modern battle. In reconnaissance it can fight to get information, fight to protect it, and fight to get it back.

Due to the importance of patrolling operations, much emphasis is placed on this subject in training for combat. It is considered a basic function of the infantry and is emphasized in unit training. A parallel can be drawn between training for patrolling and training in rifle marksmanship. All members of the infantry must be trained in patrolling just as they are trained in rifle marksmanship, as they are both essential prerequisites for combat. However, marksmen of promise are taken aside and given extra training so they can qualify as snipers. Similarly, soldiers showing particular aptitude should be given extra specialized training in scouting and patrolling for exclusive use in ground reconnaissance.

Because of the varied tasks of the infantry and its many battle missions, all patrolling actions, particularly those of reconnaissance can not always be best performed by the infantry unit, whose main interest will center around the security and screening phases. The infantry should not be expected to fight all day and patrol all night to keep the greater portion of the flow of enemy information constant.

Specially trained units must be available to the conmanding officer for the best execution of certain reconnaissance and patrolling missions. These units must receive instruction in the mechanics of ground reconnaissance far beyond that given in basic infantry training. This extra emphasis placed on scouting and patrolling by the commanding officer will be directly reflected in the successful performance of combat patrolling missions.

All ground commanders depend upon active patrolling to investigate terrain, obtain enemy information, confirm information already received, and to perform limited combat and screening operations. The closer the contact with the enemy force the more intensive the patrolling activity. Patrolling must be *continuous* in order to control No Man's Land, to maintain aggressiveness and to prevent, by any sudden increase in patrolling activity, a warning to the enemy of any impending action.

The manual defines patrolling as a detachment of troops sent out from a larger body on a mission of security, reconnaissance or combat.

Because of this difference in missions, there are three general classifications of patrols: security (screening) patrols, reconnaissance patrols and combat patrols. Within these classifications patrols can be named and formed according to the mission assigned.

The formations and procedure of patrols will vary from time to time according to the terrain and situation. Jungle, desert, and snow all create special problems, but basic patrolling principles remain the same. No attempt will be made here to cover any more than the basic precepts necessary for the success of any patrol action.

It must also be recognized that patrol fundamentals will apply equally whether the patrol is a security, combat, or a reconnaissance type. The combat patrol, although its mission is offensive and requires fighting to accomplish, must observe the same principles of security, movement, and operation as the reconnaissance patrol. The individual member of a unit making up a large combat patrol may not have achieved the same high level of proficiency as the soldier who has been especially trained for sneak reconnaissance and scouting, but the more of such training he has received and the more experienced he is, the better the chances of success for the entire patrol. The patrol is no better than the technique of its most inept member. Common sense dictates most patrol actions. The techniques laid down in manuals are generally trustworthy and mechanically correct, but they must be adapted to the terrain and local situations.

ENEMY ATTACKS

(A) *Small Patrols:* When a small type of patrol runs into an enemy ambush, the individual patrol members should disperse and make their way to a previously designated point of rendezvous or assembly. This point will be changed from time to time as the patrol progresses into hostile territory. It should be an easily recognized terrain feature or man-made object which is known to be free from enemy occupation.

If there is any possibility of accomplishing the whole or part of the mission, any member or members of the patrol must carry on after reorganization at the assembly point, unless contrary instructions have been given by the dispatching officer. Under some circumstances it may be of more value to the dispatching officer to know of the failure of a certain patrol by having its surviving members return in time for him to devise other means and missions to get the required information. *This action by surviving patrols must be taken only if previously directed by the dispatching officer.* In all other cases they must continue on their original mission.

In case of a surprise encounter with an enemy patrol at close quarters, fighting will be unavoidable and the patrol must immediately take the offensive and open fire. "He who fires first lives longest."

(B) *Combat Reconnaissance Patrol or Fighting Patrol:* Large, heavily armed patrols of this type are intended to fight off enemy attack. The principle of their use dictates the drawing of enemy fire and the location or destruction of enemy installations.

Basic small unit tactics dictate the action taken by such a patrol when it contacts the enemy. Normally the forward element of the patrol will fix the enemy with fire while the rear element brings into play its weapons on the flanks or undertakes an envelopment.

When a patrol of this type expects to encounter heavy opposition, withdrawal under smoke and covering fire usually takes place; by means of coordination with artillery, prearranged artillery fire may also be used to cover the withdrawal.

SECURITY

Patrol security is based on continuous ground and overhead observation whether in movement or at a halt. Such security for the small sneak patrol is mainly for the purpose of avoiding enemy observation and control. In the larger combat patrol, security involves the prevention of surprise contact with the enemy.

All patrol formations must make provisions for their own security. Men assigned to the point, rear, and flank act as a protective screen for the main body of the patrol. Upon them the leader is dependent for his enemy and terrain information, which will influence the actions of the patrol.

Men assigned as flank guards or acting in flank groups must maintain contact with the patrol leader. They should investigate all suspicious terrain features and possible enemy positions adjacent to or on their assigned flank.

The interval between the patrol main body or the leader and the flank guards must be flexible enough to allow these security elements to take full advantage of terrain and cover. Visibility and terrain conditions will always influence these flank intervals. Because contact with the patrol leader is mandatory there will be some occasions when connecting files between the leader and the outside flank elements are necessary.

While the patrol is in movement the flank guards move parallel to the main body and maintain as nearly as possible the same relative positions. At times the main body will have to slow down or speed up to keep the flank elements in their proper place. During the halt of a small sneak patrol the flank guards, the point, and the rear will remain in place facing outward.

In the case of a larger patrol, the same procedure will take place but additional men from the main body may be sent out to the flanks to provide the necessary all-around security.

The point and rear operate in the same manner as the flanks and must also keep continual contact with the patrol leader. Prearranged signals or connecting files can be used for this purpose.

PATROL EQUIPMENT AND ARMAMENT

(A) *The Reconnaissance Patrol:* The small (sneak) reconnaissance patrol must travel as lightly as possible. Patrol members will usually not carry a helmet, pack, gas mask, blanket, cartridge belt, or other bulky equipment unless there is a special need. They should wear tennis or other soft rubber-soled shoes to get the most out

Figure 52

PATROLS MUST MAINTAIN ALL AROUND OBSERVATION AND SECURITY AT THE HALT of silent movement. If such shoes are not available, socks worn over field shoes will work satisfactorily. Clothing should be warm and comfortable and must not interfere with movement (crouching). Garments made of soft material are better because they tear more silently and do not cause to much "scratching" noise when they come in contact with hard surfaces. Leggings should not be worn and it is best to tie the trousers around the ankle with a cord to protect the legs and to prevent from creeping. Equipment such as belts or canteens, that will cause "scratching" noises, especially when crawling, must be adjusted.

Principles of camouflage must be observed both day and night so that all clothing and exposed skin surfaces blend with the background. Equipment must be carried noiselessly and must be prepared so it does not reflect light.

The helmet or liner must not be worn and is best replaced by a knit cap. Helmets have unmistakable outlines, make noise when hit by branches or wire, prevent hearing faint sounds and distort others.

The carbine, pistol, knife, and grenade are ideal weapons for the small reconnaissance patrol or the individual scout. Naturally, the weapons available and the organization from which the patrol is selected will make a difference in armament. Special weapons for killing silently such as knives, clubs, axes, machetes, and strangle cords can be carried and are very desirable *if the men have been trained in their use.*

First aid packets, canteens, watches, knives, compasses, message pads, and binoculars are necessities for most patrols. Special equipment such as safety pins for anti-personnel mines, wire cutters, mine prodding implements and detectors can be issued for particular occasions. K-rations and tablets for chlorination of water can be issued if the mission extends over any long period of time.

Radios, flares, signal lamps, and other means of communication can be carried. Colored smoke grenades set off from prominent terrain features can be used to transmit information. (Time delay fuses could be devised which allow the user to clear the area and take cover.) Colored smoke can also be used for ground-air communication and is especially useful in jungle and desert areas. Sound powered telephones and light field wire can be used for communication with the parent unit on local missions. Judicious use of signal lamps or colored flashlights helps in some situations to solve the communication and recognition problems.

(B) *Combat (Fighting) Patrol:* Most of the equipment advocated for members of the small reconnaissance (sneak) patrol will be carried by members of a larger fighting patrol. The duration of the mission, size of the patrol, and opposition expected will necessitate changes in equipment and clothing. Armament and fire power in particular will be heavily increased. Members of this type patrol may not have to make the necessary changes in equipment and clothing (as in the sneak type patrol) which will help in concealment and silent movement. The fighting patrol members may carry full field equipment. Naturally the more emphasis on silent movement and concealment placed on each individual member of the larger patrol, the better its chances for success. However, the very size of such a patrol usually precludes silent and concealed movement such as that of a small sneak patrol.

Armament

A great amount of automatic fire power which can be easily carried in operations is a requisite of such a patrol. Sub-machine guns, automatic rifles, light machine guns, and assault rifles are carried. Grenades, both offensive and smoke, will usually be carried in quantity. The LAW rocket and grenade launcher, useful for pill boxes and anti-tank protection, will usually be part of the patrol armament. Light mortars can be carried in the support elements. For weapon dispositions in the patrol see diagrams on patrol formations.

Movement. In patrol movement the obvious or easy route should always be avoided as that is the one the enemy is most likely to cover. Patrols usually advance by bounds from one covered position to another in the same manner as the individual scout. Distances between bounds decrease as they approach known enemy installations. When approaching a dangerous locality, a scout or point man must investigate all suspicious areas while the rest of the patrol covers him. After being satisfied that it is safe to continue, the scout signals "forward", and remains in observation while the rest of the patrol advances.

The Jungle

In the jungle where the thickness of vegetation and uncertainties of terrain allow little, if any, observation, the combat-reconnaissance patrol must also be used. Here movement is ordinarily canalized along a few paths and other natural passageways. Ambushes are easy and frequent. The suddenness of enemy attacks and other types of surprise encounters make the larger combat-reconnaissance patrol a necessity. It must be strong enough to be able to overcome the enemy action or be able to withdraw behind its own protective screen and fire power.

A small reconnaissance patrol has little chance in situations of this kind. It can best be used for local purposes or in situations where it can infiltrate through jungle growth to areas where observations are sufficient to allow an accurate estimate of the actual enemy dispositions.

SELECTION OF PATROL MEMBERS

The operational success of a patrol will usually be in direct proportion to the selection of trained experienced personnel, and the number of skilled personnel it has available for this combat function.

When a basic infantry unit is called upon to furnish the bulk of patrols for operations and has only a few men who are really skilled and experienced in patrolling they will often be overworked. This is especially true if the subject has not been emphasized enough during the training period.

To use too few skilled men for the bulk of patrolling missions eventually results in casualties and in attrition of their number. Units following such a policy have found their patrolling becomes steadily weaker if no effort is made by the Commanding Officer to utilize these few skilled men more sparingly so they can be used to train others.

Because of the military importance of a patrol and because the life success of the patrol mission depends upon the individual training, personal ability, and confidence of the individual members, the patrol should represent the highest degree of training of any selected combat group. Each member of a reconnaissance patrol should be carefully selected and specially trained for this important combat mission.

Well-trained patrol members will have the needed aggressiveness and confidence for a patrol action. The individual patrol member must be amenable to discipline and preferably should understand small unit tactics in offense and defense. With this type of a background he will have the automatic reaction of the trained soldier which will help in achieving the necessary coordination to make a successful patrol.

The lone scout is responsible only for himself and his information, but in a patrol each man must not only be skilled as a scout but must also consider himself as a member of a team whose mission can only be accomplished by perfect coordination among the individual members. A violation of the basic principles of camouflage, silence, and movement by any individual patrol member may jeopardize the safety of the entire patrol.

Aside from the basic qualifications of a good scout the patrol member must be chosen on the basis of the particular mission. The members of a small sneak patrol should always be hand-picked by the patrol leader. The patrol leader must know the capabilities and limitations of his men and must have men who know and have confidence in him as a leader.

It is not advisable to pick arbitrarily an entire unit, such as a squad, for a sneak type reconnaissance patrol. The same standards of training and proficiency in silent movement, utilization of cover and concealment, and general adaptability to patrol actions will not be found in all members of the unit concerned. Such a patrol is only as strong as its weakest member.

The composition of the reconnaissance patrol should be left to the discretion of the patrol leader or the dispatching officer. Higher headquarters should only give the mission to the unit concerned and the composition of the patrol itself must be left up to the unit furnishing the patrol New men should not be assigned to important reconnaissance patrols for the purpose of gaining experience.

If patrolling is weak when a unit arrives in the battle zone, this training deficiency has to be made up. Training programs can be undertaken by a unit while in combat by sending out practice patrols for very limited distances to the front or flanks. This distance can be increased as confidence and proficiency grows. But green men should never be selected to go on an important patrol mission merely because it will give them experience.

In the case of the larger size patrol such as the combat reconnaissance type or any kind of a fighting or combat patrol an entire unit such as a squad or platoon can be selected. This method has advantages in that it will be made up of men accustomed to acting and fighting together as a unit. Such a unit patrol will react better to combat situations.

Although the method of selecting a unit for fighting patrol action has been largely satisfactory in both theaters, it is still advisable that any type patrol be made up of hand picked and specially trained men, providing they have time to rehearse as a unit.

Patrols in all cases must be made up of men who have a will to close with the enemy. There will always be some men in every unit who are not psychologically suited for patrol actions. The perpetual fear of ambush, enemy mines, and the nervous strain of action at night against an unseen enemy will be more than some men are capable of controlling. If soldiers of this type are not weeded out of the unit prior to departure on a mission, the good men of the patrol may die as a result. In case of a fire fight the good men will fight, the bad will take cover, remain motionless and think only of self-preservation.

The patrol is not a place for disciplinary action or revenge by a commanding officer and should not be made up of personnel who are assigned as a punishment for previous acts. This practice, even if it involves only one man, has an adverse effect on the concept of patrolling in general and lowers the morale of the other members.

The volunteer system (allowing men to volunteer for patrol assignment) is not generally satisfactory because the men who are actually best at patrolling will not always volunteer. A volunteer may be the type who is just looking for a fight and will do so at the first opportunity. (Such a man would not be satisfactory on a sneak patrol.) To use the volunteer system consistently as a means of selecting patrol members will also result in the belief that such missions are suicidal. Such a concept of normal routine patrolling actions will result in a lowering of patrolling standards. There will be extraordinary circumstances where volunteers may be asked for, but these should be the exception not the rule. To consistently rely on volunteers for patrol duty is in reality a confession of weakness and lack of training in a basic combat subject by members of the unit.

THE PATROL LEADER

The appointment of the patrol leader must always be made with the type of mission in mind. However, the appointment of a leader for any type of patrol is all important because his personality will reflect the ability and effort of the patrol. If the patrol leader is

Figure 53

The PATROL LEADER

THE BRAINS OF THE PATROL

THE HEART OF THE PATROL

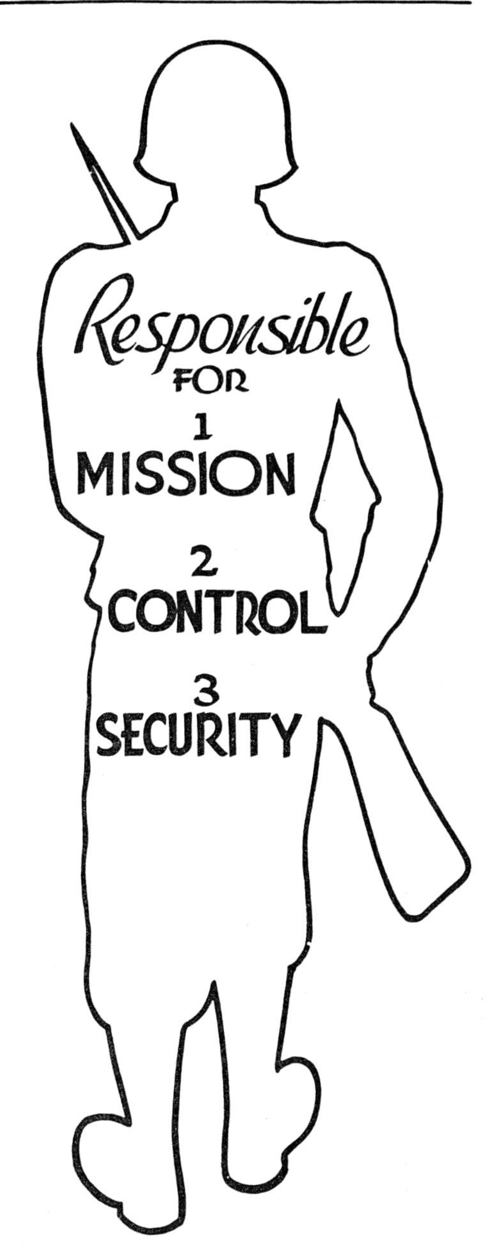

HIS POSITION IN THE FORMATION DICTATED BY THE DEMANDS OF CONTROL too aggressive and impatient, his patrol will be the same. If he is thorough and cautious, so will be the patrol. A poorly led patrol, although it may be made up of the most experienced men, has a much lessened chance of success.

Not only is it important that the patrol leader be carefully chosen, but he also must be notified of his assignment in ample time to pick the members of his patrol (if they have not already been designated), and to make necessary preparations and plans. The patrol leader may either be a skilled non-commissioned officer or an officer. This will vary among units and according to the mission and size of the patrol, but most successful patrols, particularly larger patrols of the combat-reconnaissance type, are usually led by officers.

Small sneak type patrols of two to four men may be led by reliable non-coms. If the non-coms are not trained in leading patrols and lack aggressiveness, even these patrols will have to be led by officers. It is generally true that the average officer may perform a patrol mission more aggressively as he has a greater sense of responsibility than the average non-com, but unless extra officers are available, consistent use of officer led patrols will be difficult. This is because battle casualties among company grade officers are high and line officers cannot normally be expected to perform aggressive combat duties all day and efficiently lead patrols at night. This is another reason why it is so important to have specially trained enlisted personnel for this purpose.

Combat will always develop a few soldiers who will have exceptional ability as leaders regardless of rank. These men will naturally be utilized, but such individuals cannot be expected to appear in any degree of regularity in all units.

If a soldier leads a few patrols successfully commanders will want to send him on subsequent missions. This is a normal procedure as long as it is not abused. There is a tendency by some commanders to "work a good horse to death". One or two good patrol leaders (officers, non-coms) should not be expected to lead all the unit's patrol missions. Eventually they will become casualties and their value and experience will be lost to the commander, who will have to start all over again by picking men of unknown ability to lead subsequent missions. Again this is usually the result of a lack of complete unit training in patrolling and patrol leading.

In the early actions of WW II, often too great a proportion of small reconnaissance patrols had to be led by officers. This was a result of insufficient training in scouting and patrolling and a consequent reluctance to place responsibility in the hands of the non-commissioned officers and enlisted men by the dispatching officer. At times, intelligence officers found themselves doing all the forward reconnaissance because of a lack of men in their sections who could do the job. Replacements of these intelligence officers were often very frequent due to casualties which would not have occurred if they or responsible noncommissioned officers had been well-trained for the task of leading the patrol. Occasionally officers will be designated to lead small reconnaissance (sneak) patrols of special importance but normally they should be led by good reliable non-commissioned officers.

If a complete unit is designated as a patrol, the leader of that unit should normally be the patrol leader. Again, the mission and its importance must be weighed against the disadvantages of a unit patrol led by someone other than its normal commander. The situation may be such that a special officer will be assigned to lead the unit. This will be a command decision and the unit leader will normally be second in command.

SECOND IN COMMAND

In choosing his second-in-command, the patrol leader must be certain that the man is capable of taking over the patrol and completing the mission if the patrol leader should become a casualty. Such a man should possess, as nearly as possible, the same leadership qualities as the patrol leader. It is the responsibility of the patrol leader to thoroughly brief his second-in-command and to be sure that he has a complete knowledge of the mission. The sooner the second-in-command is designated and briefed the better he will perform his job. Usually when an entire unit is chosen for the mission, and is led by its regular commander, the second-in-command of that unit will be the second-in-command of the patrol.

If the patrol is to be of large size it is advisable for the second-in-command to be present when the patrol leader is being briefed by the dispatching officer. If possible, the patrol leader should take the second-in-command with him when he makes any observation of the terrain prior to the mission. Prior to departure on the mission the second-in-command should be assigned tasks which will relieve the patrol leader of some of the burden of preparing the patrol.

The point man of any patrol (scout) should be the most observant and experienced of the patrol members. He must be thoroughly familiar with the compass, use of maps and terrain in general. His ability at maintaining direction, using cover and concealment, and other skills, should be a known and proven factor. A knowledge of the tactical employment of weapons in respect to terrain and of enemy habits and ambush techniques will make him better able to investigate and evaluate any suspicious areas before the patrol is committed.

PATROL SIZE

Because of the varied types of patrol action, and dependence on the particular situation, three factors always exist which determine patrol size. They are:

(1) *The enemy* - The racial and fighting characteristics of the enemy and knowledge of his counter-reconnaissance methods, tactics, strength, weaknesses, methods of deception and information gained from previous contacts with him will help to determine the size of the patrol. In other words, *Know your enemy.*

Figure 54. The duties of a patrol member.

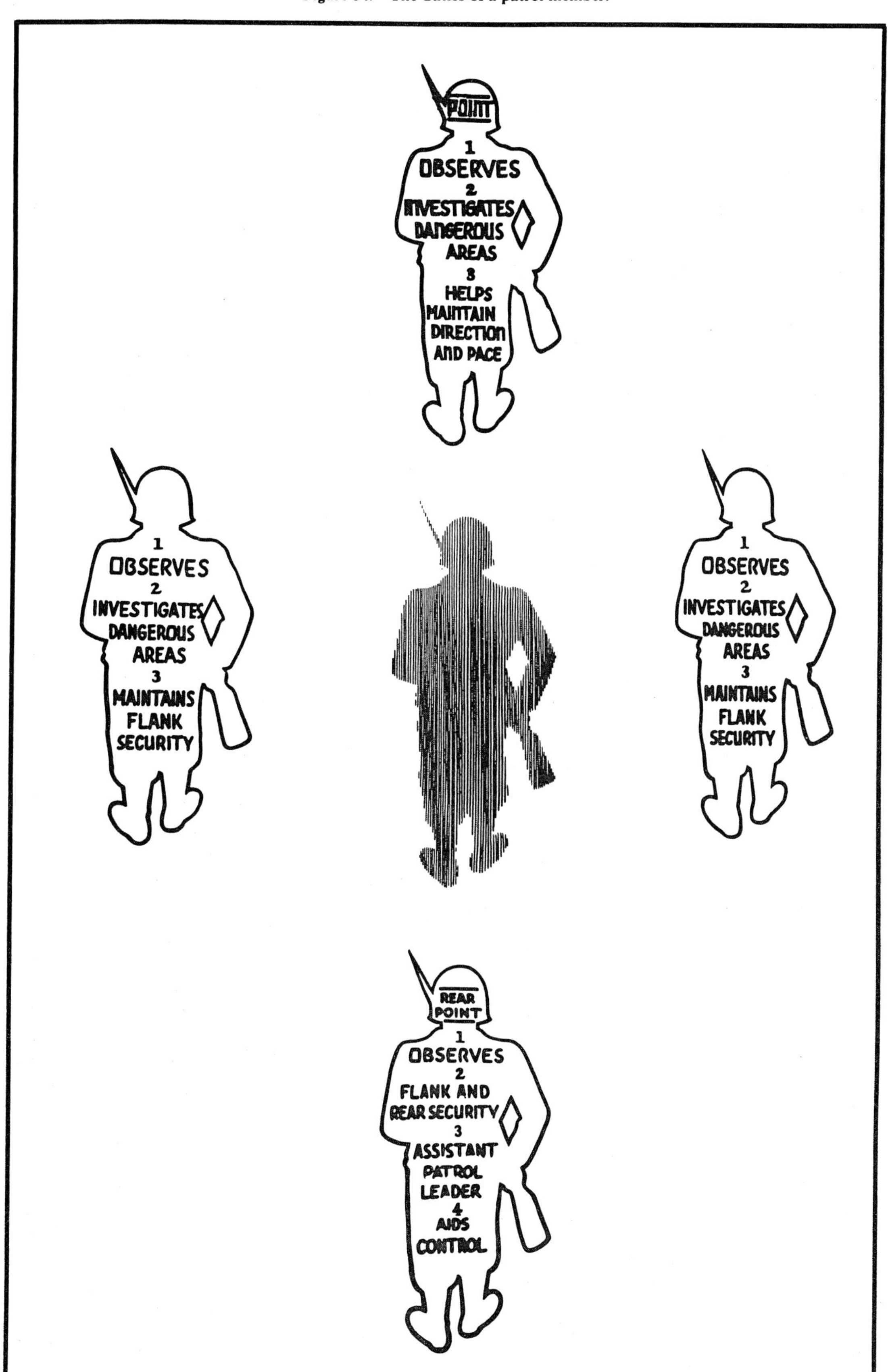

(2) *Terrain* - Affects the scope and character of all military operations. The formation of the ground between the patrol and its objective and the amount of natural cover on it will always be determining factors in patrol size.

(3) *Mission* - Patrol size in respect to the mission will generally be affected by the purpose for which it is formed.

These three questions coupled with the situation will help determine strength:

Is it a mission involving stealth (day - night)?

Is it a mission involving fighting (day - night)?

Is it a mission involving both stealth and fighting?

(A) *Reconnaissance* (Sneak)

Because stealth is the means by which the small sneak reconnaissance patrols accomplish their mission, silent movement, security and concealment are the limiting factors on size. Such a patrol basically should consist of three men.

A three man patrol is the minimum number that can provide all around observation and security for itself. Such a patrol team is easy to control and can maintain the necessary balance between speed and silence.

Although a three man team is the best size for a sneak patrol such patrols can be larger, but increases in size make them harder to control, harder to conceal, slower in progress and increase the difficulties of silent movement.

If a local mission is analyzed carefully it will usually be found that the three man patrol can do the same job that five or six man patrols can. Sneak patrol missions requiring messengers to be sent back or those operating over great distances may have to be larger. But reconnaissance patrols used in normal missions which number five and six men are in reality **neither "fish nor fowl"** being too large to get the greatest benefit out of silent movement, cover and concealment and in most situations too small and lightly armed to fight for information or fight to get it back.

(B) *Combat Reconnaissance Patrols - Combat Fighting Patrols*

As previously stated the combat reconnaissance patrol and the pure combat fighting patrol can be identical in factors such as size, armament and formations. The difference lies in *the primary mission.*

Such patrols may vary from one officer and a few men to a company or more. This wide variance in size is dependent on the opposition expected, nearness of contact with the enemy, and the other basic influences in patrol size which can only be determined by the situation. On a fluid front where lines are not definitely established the large heavily armed patrol can be sent ahead to an area where detailed reconnaissance is needed and there it will act as a base from which small sneak patrols are sent out. Such a patrol operating well ahead of the main body will have to possess sufficient strength to be able to send out these smaller patrols and still have the power and numbers needed to fight off any enemy action.

It is generally true that the farther a patrol operates ahead of its own troops the stronger it must be. However, this should not be an iron clad policy as frequently, a number of well-armed small patrols of less than squad size can penetrate hostile country and infiltrate enemy lines better than the larger ones.

The best criterion for patrol strength is always gained by profiting from the experience of other units under similar circumstances against the same enemy. In operation on a newly established front, the commander should use basic patrol concepts initially and make any variations from this to meet the immediate needs after he has had a chance to evaluate the success or failure of his patrols, and those of similar units.

FORMATION

Patrol formations are many and varied because they are subject to many influencing factors such as: terrain, strength, mission and enemy opposition. During the mission the patrol may have to assume different types of formations to meet changes in these influencing factors.

A good well-trained patrol will change formations automatically to meet variations in terrain and cover.

There is however, one governing principle which must be constant at all times: *the maintenance of control by the patrol leader.* Without control there ceases to be a formation and the patrol becomes a disorganized group, operating in the face of the enemy. Day and night are also two great influencing factors on patrol formations. In the daytime when visibility is good, contact and control are facilitated; therefore, intervals between patrol members must be reduced to meet the demands of control.

An inexperienced unit will obtain better results if it uses simple formations and limits the size of its patrol until the standard of training and experience is such that the difficulties of control are overcome.

There are two basic foundations around which any patrol formation can be developed.

(a) *The diamond* is a formation where the majority of the strength of the patrol is distributed around the four points of the diamond.

Such a patrol formation provides all around security and observation in movement, as well as the halt. Good control is possible provided the patrol strength is not too great. This formation, with variations, can be used to best advantage in daylight, in open terrain and thinly wooded areas where observation is good.

(b) *The column* is a formation where the main portion, or the strength of the patrol, is echeloned in files behind the point and leader of the patrol. Men are placed on either flank to give all around protection and observation while the patrol is in movement. At the halt, all-around security is obtained by having the flank men remain in position. they may be strengthened from the main column if the situation warrants.

Figure 55

Basic
DIAMOND FORMATION

Provides all around Security and Observation in Movement and at a halt

Best Employed,

1 OPEN TERRAIN
2 BRIGHT NIGHTS
3 THINLY WOODED AREAS

Figure 56

Basic
COLUMN FORMATION

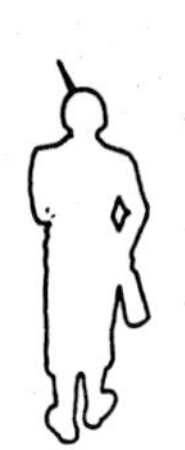

INTERVAL

Governed by Visibility, Terrain, and demands of Control.

FLANK PROTECTION

Governed by Terrain, Visibiliy, and demands of Security.

1 DARK NIGHTS
2 DENSE UNDERGROWTH
3 BAD WEATHER
4 WHEN TRAIL BOUND OR CANALIZED

This type of formation is often called a T variation of the diamond but in reality it is different from the diamond because the strength of the patrol is in the center of the column.

The column formation is best used in conditions where visibility and control are poor because of light and weather (fog) conditions or dense undergrowth (jungle). It is usually the only formation possible when movement is restricted to jungle or mountain trails or in deep snow.

It is often inadvisable to use the diamond formation at night, because it is very difficult for the flank men to maintain a proper position in relation to the main body when they have to converge and diverge according to the terrain. In dense growth or on dark nights, the flank men will cause an increase in noise, get lost, lag behind, get ahead, or fail to respond to the commands of the patrol leader.

Adverse conditions may dictate that a column formation without any flank position will have to be used. At times physical contact between members of the column will be necessary to maintain direction and control. At other times, luminous markers placed on the back of men in the column have been used for this purpose. Under more favorable conditions intervals between men in the column formation will be dependent on the same factors as those influencing intervals in the diamond.

The position of the patrol leader: Neither patrols or men can be led from the rear. For this reason the patrol leader must always be near the forward elements or point of his patrol. His position, other than the above limitations, must not be a fixed one but should be flexible enough so that he can maintain control and make the necessary observations and decisions before committing his patrol to any type of action or terrain obstacles.

In the small patrol the second-in-command should be either in the center or rear of the patrol. He should not be too close to the patrol leader as an ambush might result in both being casualties. In the larger patrol he would normally be in command of any support element. In any case, he should be in a position to help maintain control.

Means of control: If a patrol is made up of experienced men who have learned through training and rehearsals to operate as a team, the problem of control will be simplified. If at all possible, it is desirable for the patrol leader to know the names of his men. In emergencies and fire fights he can control them easier.

Small patrols during daytime can be controlled by oral orders and prescribed arm signals, whereas similar night patrols are controlled by prearranged sound signals. The signals for the night patrol may imitate natural sounds, such as bird calls, or dogs barking, or those which carry only short distances, such as scratching the fingernail on a match box, tapping or rubbing the palms together. Such signals sould always be tested during rehearsals for suitability. The signals should always include "Halt", "Forward", and "Check-up". Any member of the patrol may give the signal to halt. The member who halts the patrol is normally required to start it again, unless the leader has specified otherwise. Only the leader should give the check-up signal. Signals must always be predetermined and covered by the patrol leader prior to the patrol mission.

In larger patrols, arm, hand and sound signals may also be used, but they are not always sufficient for good control. Runners, connecting files, flares, whistle blasts, and radios may be utilized. Such additional means of communication and control within the patrol are particularly useful when the formation consists of a forward and support element separated by a terrain feature.

Control from parent unit: On a front that is strongly held, it is often difficult to control the actions of the numberous friendly night patrols that will encounter each other. in the darkness where faulty recognition signals and other factors can often cause friendly patrols to fire on each other. Occurrences of this type create a lack of aggressiveness and confidence in the actions of the individual patrols.

Any means which can be devised to facilitate recognition and avoid fire fights between friendly patrols operating in the same vicinity should be used. Radio communications with the unit sending out the patrols will help solve this problem. Flares and signal lamps can also be used, security permitting.

On some occasions, friendly patrols operating at night in a sector where occasional contacts with each other are probable and may result in fire fights can use the sound power telephone with light field wire to maintain contact and control with the parent unit. Each patrol operating in the sector will lay wire as it progresses forward, checking in at specified intervals, with the central switchboard on which all patrol wire is connected. By this means, the dispatching officer can direct patrol action, definitely locate them, and by using the board can connect one patrol with the other. Accurate, timely, patrol operation and control is assured. Patrols returning before dawn can do so by following and picking up their own wire. However, danger of enemy ambush when returning to pick up the wire must be considered.

The psychological effect of the patrol being in constant touch with its unit also has a great effect on the aggressiveness and actions of the individual patrol. Green patrols in particular are benefited by the use of sound power phones in this manner.

Security Patrols

The main mission of the security patrol is to build up a counter-reconnaissance screen or prevent surprise attacks by the enemy. This is done by the basic infantry unit and once lines or contact are established it usually consists of patrolling action to the immediate front and flanks.

Security patrols not only perform a screening mission against enemy reconnaissance units by preventing penetration of friendly lines by hostile patrols, but they also operate on the flanks maintaining contact with friendly units and covering the weak spots between adjacent units which are always subject to probing and attack by the enemy. In some cases the screening patrol has replaced the concept of the positional out-guard and in addition to accomplishing the mission of preventing enemy intelligence of friendly positions they are used operationally to control No Man's Land. This is particularly important to an army on the offensive.

The size and composition of the security type patrol will always depend on terrain and situation. It must be very flexible to meet the needs of changing combat situations. Because the nature of security missions often necessitates fighting, combat patrols are often used for this purpose.

Reconnaissance Patrols

A reconnaissance patrol may be used to secure enemy information, maintain contact with the enemy or procure information of the terrain. There are two types of reconnaissance patrols, the small "sneak" patrol and the larger combat-reconnaissance patrol.

The sneak patrol avoids unnecessary combat and accomplishes its mission by stealth. It is lightly armed and fights only in self-defense. Locations of enemy positions and tactical dispositions are gained by close observation only. Such a patrol is normally composed of about three men and seldom exceeds six men in size. It is usually led by a commissioned or non-commissioned officer. Many times sneak patrols are sent out prior to larger patrol actions.

The combat-reconnaissance patrol is nothing more than a small version of reconnaissance in force. It is large enough and carries sufficient fire power to fight small actions and to protect itself when subjected to enemy fire and ambush. It can penetrate thinly held counter-reconnaissance screens thus forcing the enemy to disclose his main positions by providing enough of a threat or offering a large enough target to draw fire from main enemy installations.

With the exception of the mission, the size, armament, and formation of the combat-reconnaissance patrol are much the same as those of the combat patrol. A combat reconnaissance patrol has as its primary objective the securing of enemy information. It is strong enough to fight to get it, fight to retain it, and fight to bring it back from hostile territory. Often the sneak patrol is too weak to perform such a mission. This is particularly true when observation is limited and enemy camouflage is good, when ambushes are frequent, or when contact is so sudden that a fire fight is unavoidable and often necessary to get the information back.

In WW II combat against the Germans it was found that they would often let small reconnaissance patrols pass in through their positions and not fire upon them or otherwise disclose their positions. If the patrol penetrated too far, it would just disappear. The only way the Germans could be forced to disclose their positions, was to present a good target to them. Combat reconnaissance patrols of platoon strength or greater had to be sent out to draw fire, so that enemy positions could be located.

Another WW II example of this was in Africa, where an American regiment was scheduled for an attack. Many sneak reconnaissance patrols were sent out, and all reported that they penetrated within a few hundred yards of the objective and saw little enemy activity and only a few enemy soldiers. The obvious conclusion was that the area was lightly held. High headquarters also accepted at face value these reports of sneak reconnaissance, and concluded that their objective was lightly defended.

In the attack the next morning the opposition was so heavy that what had been considered an easy objective took eleven days to take. The Germans had taken good concealment measures and let the various sneak reconnaissance patrols pass through and return with false estimates. A combat reconnaissance patrol would have forced the development of a fire fight which would have given an indication of the true strength of the German position.

Any defensive system in depth with a screen of lightly held positions out in front can result in a false estimate of the true location of the main line of resistance if only a small "sneak" type patrol is used. Such a patrol depending only on observation might return with information of enemy installations and dispositions in the OPL which may be construed to be the principal enemy positions. An attack acting on this supposition may result in decimation of the attacking force by prepared machine gun and mortar fire from positions in the M.L.R. The combat reconnaissance patrol will many times penetrate this screen in front of the M.L.R. by attacking and enveloping any light resistance encountered or forcing the withdrawal of these forward units and bringing down on itself fire from enemy positions in the main defense line. This gives a truer picture of the location of the main enemy positions. Such a patrol is usually too large and powerful for the enemy to let pass through prepared fields of fire. Its size and armament force the enemy to open fire. It must be remembered that the enemy may be in doubt as to whether or not the patrol is the forerunner of an immediate attack and thus, the enemy cannot take the chance of letting it penetrate his lines too far.

The Combat (Fighting) Patrol

A combat patrol undertakes missions which are offensive in nature requiring fighting to accomplish. It is armed and organized so it can operate by force and has as its primary objective the harassing and destruction of the enemy and his installations. A more precise designation would be to call it a fighting patrol.

Such patrols may engage in missions such as the raiding of enemy supply points, ammunition dumps, and command posts. They can also be used to capture prisoners, cut enemy communications, and ambush enemy patrols. Fighting patrols are often assigned the task of taking and holding limited objectives as part of a coordinated infantry attack or to secure a line of departure.

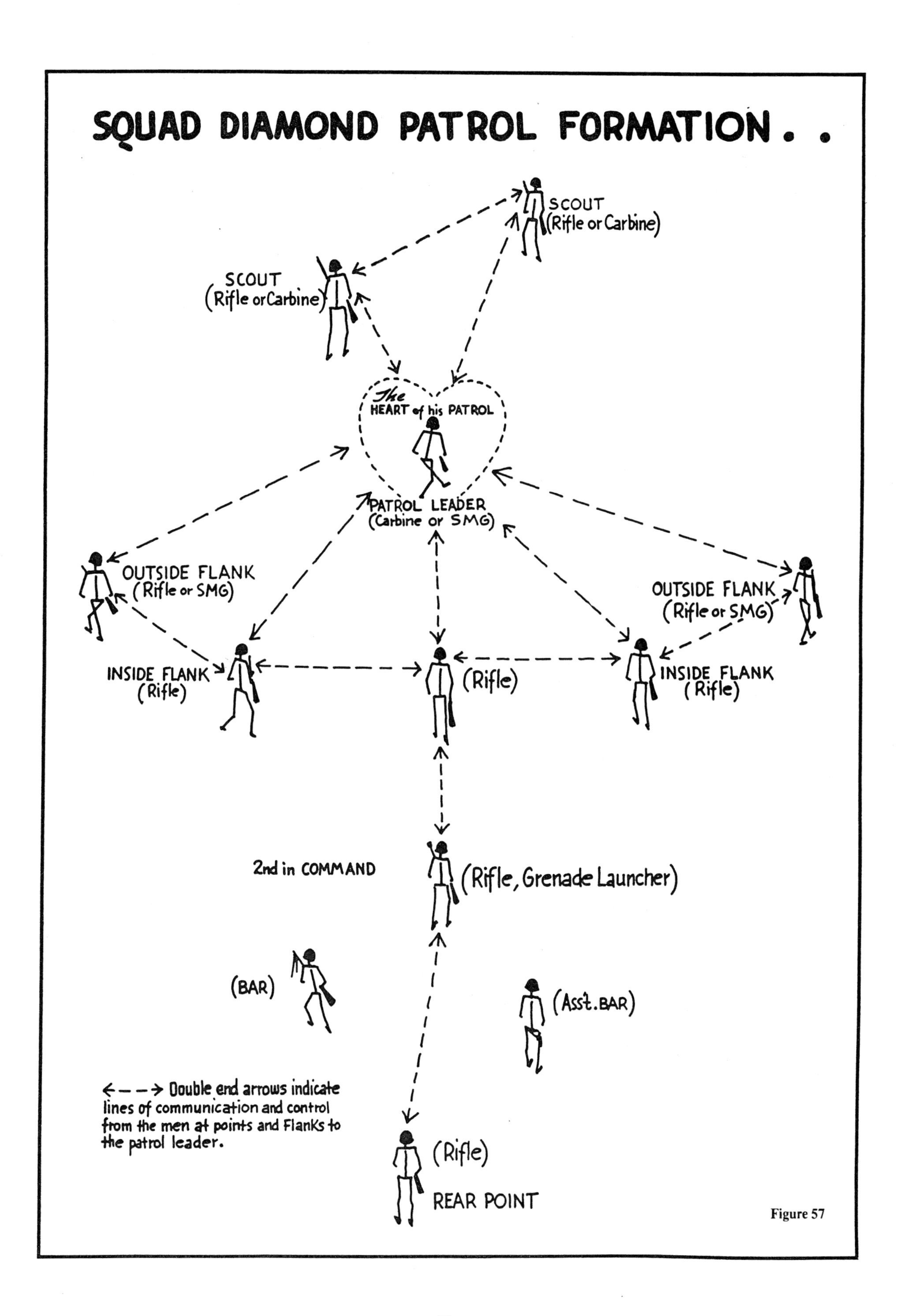
SQUAD DIAMOND PATROL FORMATION . .
SCOUT
(Rifle or Carbine)
SCOUT
(Rifle or Carbine)
The HEART of his PATROL
PATROL LEADER
(Carbine or SMG)
OUTSIDE FLANK
(Rifle or SMG)
OUTSIDE FLANK
(Rifle or SMG)
INSIDE FLANK
(Rifle)
(Rifle)
INSIDE FLANK
(Rifle)
2nd in COMMAND
(Rifle, Grenade Launcher)
(BAR)
(Ass't. BAR)
Double end arrows indicate lines of communication and control from the men at points and Flanks to the patrol leader.
(Rifle)
REAR POINT
Figure 57

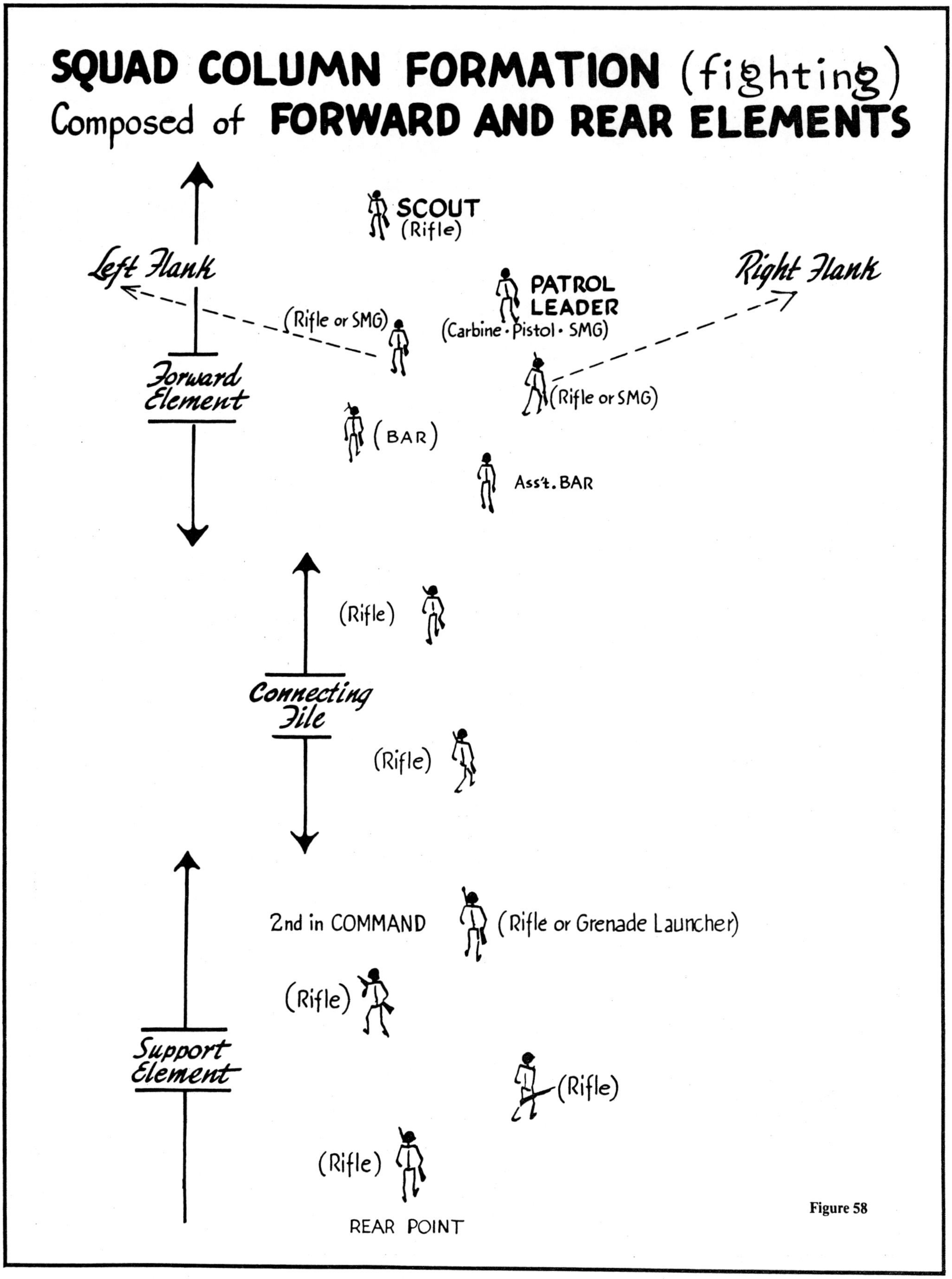

SQUAD COLUMN FORMATION (fighting)
Composed of FORWARD AND REAR ELEMENTS

SCOUT
(Rifle)

Left Flank

Right Flank

PATROL LEADER
(Carbine · Pistol · SMG)

(Rifle or SMG)

(Rifle or SMG)

Forward Element

(BAR)

Ass't. BAR

(Rifle)

Connecting File

(Rifle)

2nd in COMMAND
(Rifle or Grenade Launcher)

(Rifle)

Support Element

(Rifle)

(Rifle)

REAR POINT

Figure 58

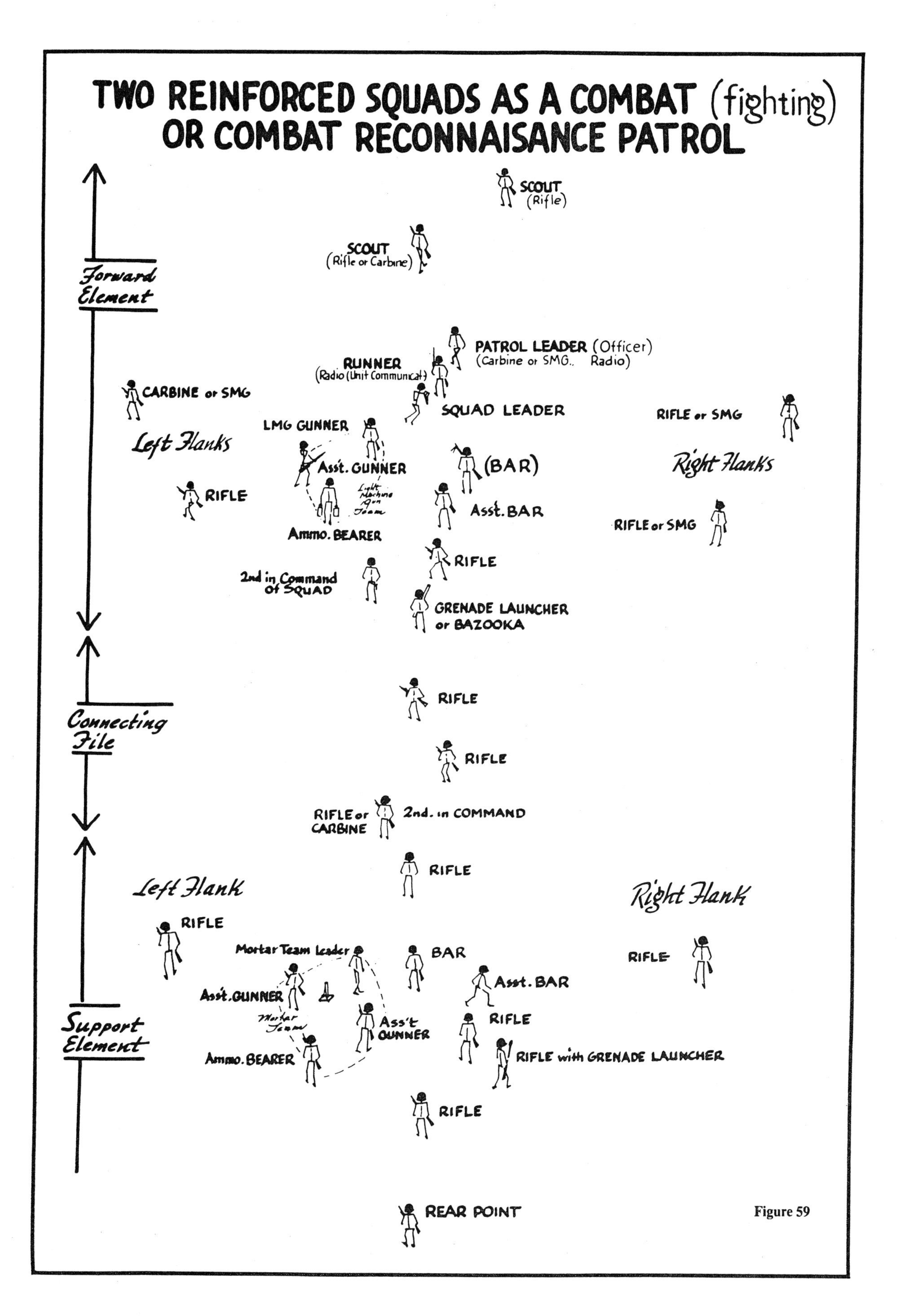
TWO REINFORCED SQUADS AS A COMBAT (fighting) OR COMBAT RECONNAISANCE PATROL
SCOUT (Rifle)
SCOUT (Rifle or Carbine)
Forward Element
PATROL LEADER (Officer) (Carbine or SMG.. Radio)
RUNNER (Radio (Unit Communicat)
CARBINE or SMG
SQUAD LEADER
RIFLE or SMG
LMG GUNNER
Left Flanks
Asst. GUNNER
(BAR)
Right Flanks
RIFLE
Asst. BAR
RIFLE or SMG
Ammo. BEARER
RIFLE
2nd in Command of SQUAD
GRENADE LAUNCHER or BAZOOKA
RIFLE
Connecting File
RIFLE
RIFLE or CARBINE
2nd. in COMMAND
RIFLE
Left Flank
Right Flank
RIFLE
Mortar Team Leader
BAR
RIFLE
Asst. BAR
Asst. GUNNER
Mortar Team
Ass't GUNNER
RIFLE
Support Element
Ammo. BEARER
RIFLE with GRENADE LAUNCHER
RIFLE
REAR POINT
Figure 59

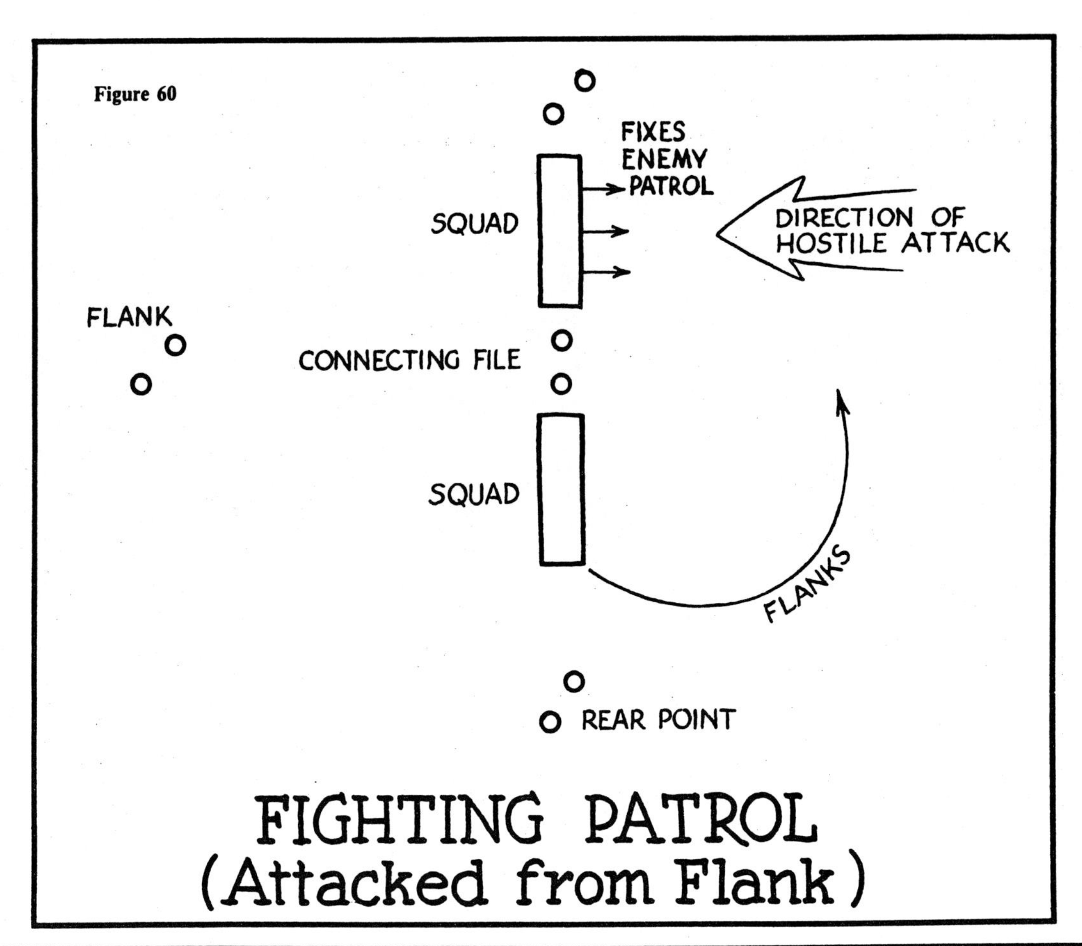
Figure 60
FIXES
ENEMY
PATROL
SQUAD
DIRECTION OF
HOSTILE ATTACK
FLANK
CONNECTING FILE
SQUAD
FLANKS
REAR POINT
FIGHTING PATROL
(Attacked from Flank)

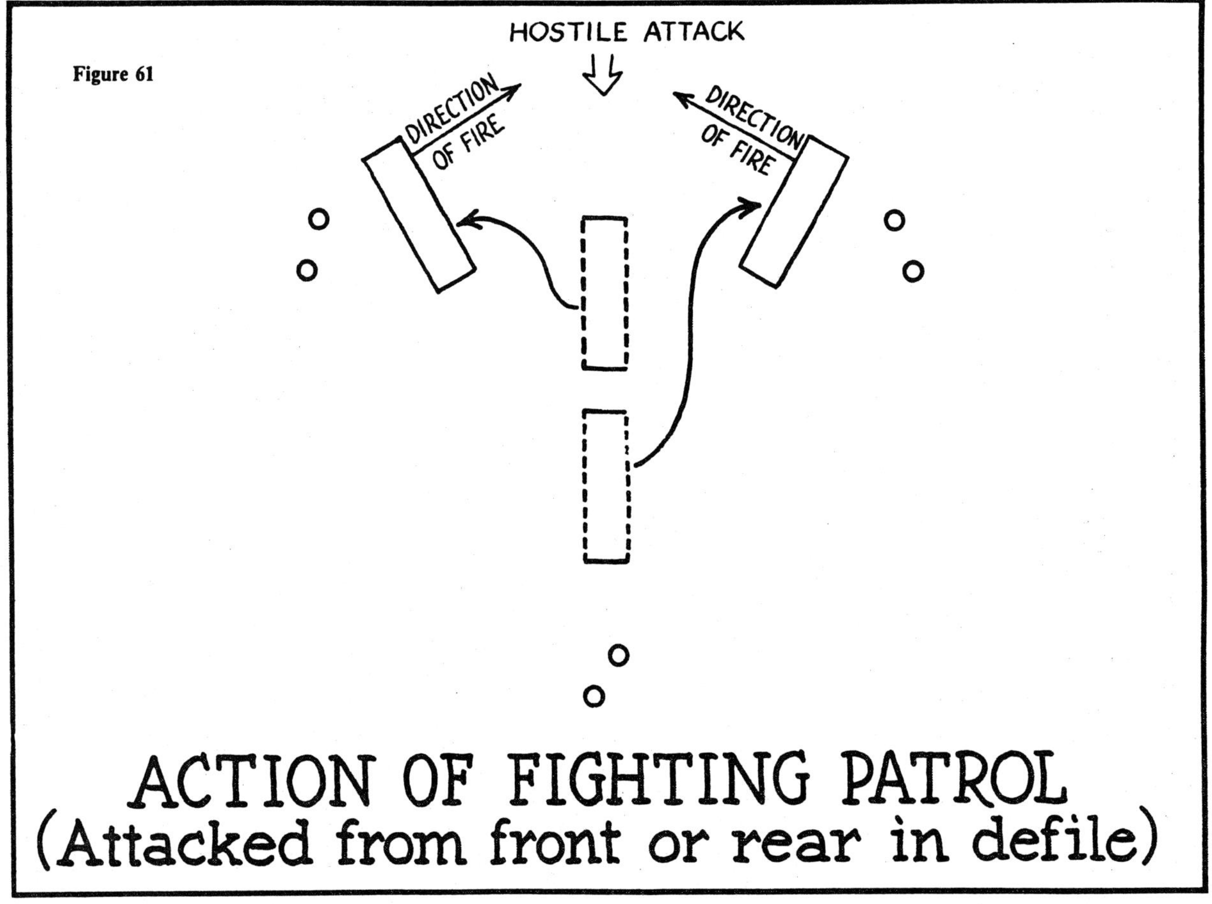
HOSTILE ATTACK
Figure 61
DIRECTION
OF FIRE
DIRECTION
OF FIRE
ACTION OF FIGHTING PATROL
(Attacked from front or rear in defile)

Defensively, combat (fighting) patrols can be used to deny commanding ground permitting observation to the enemy, to protect routes of supply, conduct rear guard actions, and to mop up infiltrating enemy patrols, snipers or airborne troops. A combat patrol can be used to assist in screening a withdrawal and to conduct demonstrations of activity designed to deceive the enemy. Fighting patrols are used to harass the enemy, control No Man's Land, and to build up confidence and maintain aggressiveness during close-contact, static situations.

The value of the use of fighting patrols to control No Man's Land can best be brought out by the following WW II example.

On the Anzio beachhead one division relieved another in an active sector. The division relieved had not done any intensive patrolling and as a result of this the German lines moved closer and closer until they were with in 150 yards of the American O.P.L. This resulted in their light automatic weapons becoming a constant behind-the-line threat to American troop movements, supply services, and a definite hindrance to assembly of troops prior to launching an attack. The division taking over the sector instituted a three week program of intensive, continuous patrolling. Numerous combat patrols of various sizes were used for harassing and elimination of German forward positions. This patrolling was so intensive and successful that the Germans were forced to pull back their O.P.L.'s and the effectiveness of their small arms fire over and behind the American lines became very limited.

In offensive situations combat patrols are used to destroy by-passed enemy pockets of resistance or wipe out emplacements and other enemy disposition holding up the advance of the main body. When the enemy is in the retreat, such patrols are used for harassing and to maintain contact.

Any patrol missions not having as their prime objective the procuring of enemy information can be classified as a type of combat patrol. The differentiation between security patrols and combat patrols used defensively is not great, but for purposes of clarity, it is better to use the classifications stated above.

It must be remembered that any combat patrol will have as a secondary mission the securing of enemy information. In fact certain specialists may be assigned to such a patrol to procure any pertinent enemy information valuable to intelligence which may develop from the action. On patrol actions necessitating operations a long distance from the parent unit, a combat patrol may become the base for smaller reconnaissance patrols sent out from its personnel.

Observation is continuous. The principle of all-around security at the halts as well as in movement should never be violated. When approaching any area where the patrol is forced to halt and wait while the point or scout investigates, "Bunching" must be avoided. "Bunching up" prior to going over or through any obstacle violates security. Enemy ambushes are often laid at obstacles of this type so as to take advantage of any failure to retain proper patrol security and formation.

Daylight patrols must utilize all types of wooded terrain and defilade areas for concealment. Although such covered areas limit enemy observation, they also provide excellent opportunities for ambush. Formations, control and security measures must be adopted to suit the amount of natural growth or defilade. Generally, intervals and distances will be telescoped to meet the density of woods and other natural growth.

Before a stream is crossed, the opposite bank should be carefully observed. The scout crosses first, the other members cover him, then follow him individually. Observation is maintained from both sides of the stream until the last man crosses. It is safest to cross a stream at bends because enemy vision up and down stream is restricted.

On approaching a crossroad or intersection, a patrol halts and sends scouts to the flanks to reconnoiter the side roads. The patrol leader then sends a scout ahead, who is in turn covered by the rest of the patrol. After he has signalled "forward", the rest of the patrol advances.

When approaching a crest or skyline, the patrol takes up a covering position from one to two hundred yards distant while the scout cautiously moves to a position on the crest where the skyline is broken. If everything is clear, the scout drops back until he is clear of the skyline and signals "forward". The patrol then approaches in a deployed formation, leaving at least one man back under cover until the others have passed over the skyline.

Gullies, defiles, and narrow ravines are danger points for ambush. When passage through such terrain obstacles running parallel to the advance is unavoidable, special precautions must be taken. The formation can be split, each section keeping up on the sides of the defile and covering the side of the defile opposite, or, if the defile is too wide, it may be advisable to keep the patrol intact and move up one side only. Caution against getting up too high on the side of the defile must be exercised. Silhouettes can be visible on the sides of slopes particularly if the enemy is on lower ground. If the length of the defile is short, scouts can be sent through to the other end before the bulk of the patrol advances.

If passing through inhabited localities is unavoidable, patrols should circumvent main thoroughfares and thickly populated districts. When proceeding up a street, a staggered formation preceded by scouts should be maintained. Men on each side of the street, next to the buildings, will cover windows, roof tops, and door ways on the opposite side.

In most cases, the investigation of suspicious areas will automatically be duties of the point or scout. On signal from the scout, the patrol leader may leave the formation and work his way to the scout for observation and to make decisions affecting the future progress of the patrol.

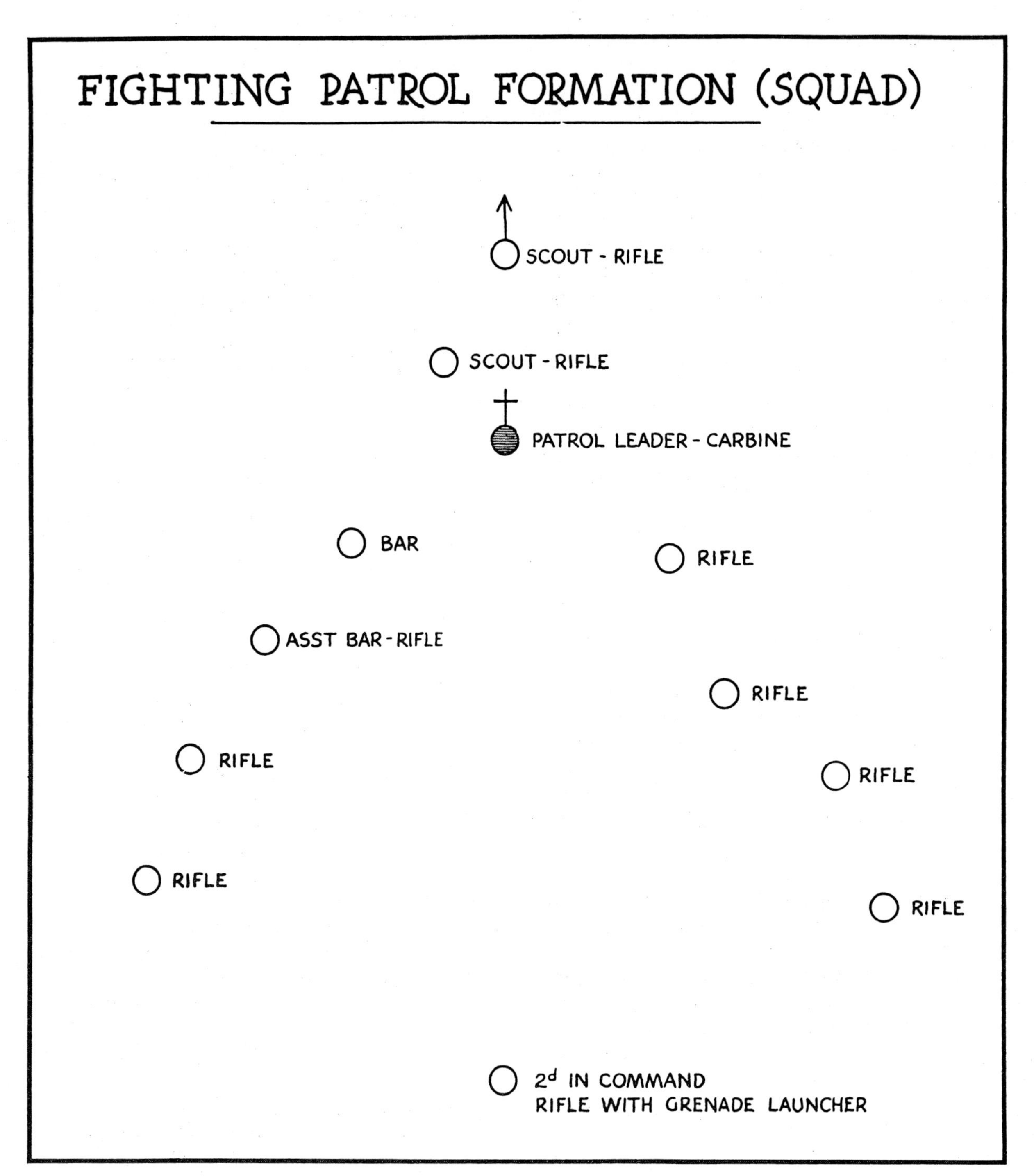

Figure 62. This "V" formation is useful for clearing areas of isolated enemy snipers and opposition. The wings of the V may lie extended or contracted, as the cover and terrain dictate.

Halts during the patrol movement are necessary for listening, observation, and rest. Such halts should be made in covered areas where all-around defense and observation can be maintained both day and night.

THE NIGHT PATROL

The night patrol has recently become of more importance than the day patrol. In any fairly stabilized front, there is little chance for operation of patrols in daylight. Enemy observation, ground and aerial and counter reconnaissance measures are usually too good.

Night patrolling (which must be practiced intensively) is one of the most important infantry operations of modern warfare. In fact, night patrolling operations are the principal medium by which a great many actions are won. The governing factor in successful night patrolling is control. Good control is based on simplicity in formations, signals, silent movement, and rehearsals.

Aside from the protection afforded by darkness, night reconnaissance patrols have certain advantages over daylight patrols in gaining accurate information. During the night, the enemy is usually making the necessary movements to establish his lines or change his dispositions for the following day. Thus, many times, information gained by daylight patrols is obsolete before it can be used. Enemy sentries are often less alert at night than in daytime, and the darkness allows night patrols to get nearer enemy positions.

The leader of a night patrol, his second-in-command, and as many of the patrol members as possible, must be given every opportunity to observe the terrain to be covered during the daylight hours. Scouting and patrolling operations will be more successful at night when the terrain to be traversed has been studied during the day, and when all available aids such as aerial photos and maps have been utilized.

Before starting, the patrol leader plans his route, selecting bounds and the locations of prominent terrain objects which will help him maintain direction. In planning the route, he should avoid sound traps and other terrain obstacles which hinder movement. Formations may be similar to those used during daylight, but distances and intervals will be reduced. Patrol leaders must pick out landmarks which can be identified at night. They must plan the route carefully utilizing "dead sound zones" or open spaces and lower elevations, to keep the enemy above the patrol. Every man should know his place in the formation, how he should proceed, what the mission is, and what to do in emergencies. The use of stream beds and ditches reduces the patrol's silhouette (being ever mindful of mines and booby traps) and aids in concealment. Every time a man makes a noise, the entire patrol must stop and listen until the patrol leader signals "advance". Whenever possible, the noises of nature and battle should be used to cover the advance.

A diamond security formation at the night halt is advisable. The diamond patrol formation for movement may be satisfactory in open areas on a clear night, but in the jungle, dense woods, on dark nights or on bad terrain, in fact, in almost all night work, single file is the only formation that allows the patrol leader to control his patrol, keep contact with his men, and maintain proper direction.

In battle zones stripped of growth and when contact with the enemy is imminent, the diamond formation can be used for greater security. Large patrols should not be split into various parties, each proceeding separately, except in favorable terrain and for short bounds. The danger that units may cross each other and open fire must be guarded against since such accidents have happened to patrols of all nations. Control of the patrol and silent movement are all important at night. Some of the differences between day and night patrolling are:

(1) Night patrols move by shorter bounds than daylight patrols.

(2) When possible, night patrols should use a more direct route over open ground or along well-defined lines such as roads, railroads, streams, and paths rather than the indirect, covered route a daytime patrol must ordinarily take. This is necessary because of the difficulties in recognizing landmarks at night. It is easy to get lost or confused. Movement over broken or thickly wooded ground should be avoided as such areas are sound traps.

(3) Night formations are similar to those used by daylight patrols, but the interval between members is usually closer so that control can be maintained by the patrol leader. Large patrols may advance in double file protected by scounts stationed to the flanks and rear. One of the most successful night reconnaissance patrol formations is, however, *single file* with the men as close together as is necessary to maintain contact. When in dense and dark woods, the men in column may have to maintain actual physical contact to avoid being separated. At night or in deep woods, a most satisfactory way to maintain contact is to make the man in front responsible for contact with the man behind him.

Good night patrolling results from extensive night training. Training exercises should include:

(1) Problems to extend from daylight to midnight.

(2) Problems to extend from night to noon.

Since landmarks appear different at night, training for night patrolling should initially be practiced on known terrain in daylight, and later over the same terrain at night in order to emphasize the difference in the appearance of the same objects during daylight and darkness. The training can then progress until all night exercises are successfully conducted over strange terrain which has been studied during daylight hours.

Figure 63. When patrolling dense woods on a dark night, a column should be used. Physical contact between men should be used if necessary to maintain control.

REHEARSALS

Irrespective of how careful preparation of the patrol has been, the factor of the unforeseen is always present. Luck plays a part in patrol activity, but careful and extensive preparation and rehearsals are prime requisites. A poorly trained, haphazardly selected and unrehearsed patrol may be lucky once, but seldom twice. Failure or success depends upon the way in which individual patrol members respond to the demands of the patrol leader, the terrain, and the unforeseen.

Rehearsals are vitally necessary for the success of a patrol mission. These necessary rehearsals which make the patrol function like any well-trained team must be conducted if time and other circumstances permit. Patrol missions and orders must always be decided upon and issued so as to allow time for this necessary preliminary. For all special patrol operations such as raids, the rapid successful execution of the missions is usually an indication of previous rehearsals.

Patrol members who have worked together repeatedly will not need to receive the same amount of stress on rehearsals as the green or newly formed patrol, but whenever a mission is out of the ordinary, a rehearsal must be scheduled.

New patrol members must always be taken out and rehearsed with old members. The following points should be covered in any rehearsal:

1. Changes in formations to be adopted automatically when encountering different terrain obstacles (day and night) and the procedure to be used in traversing them.

2. Action to be taken on surprise encounters with enemy elements.

3. Action to be taken if ambushed during a withdrawal.

4. Signals and other means of maintaining control.

5. Responsibility of individual patrol members and methods for maintenance of security (movement and halts).

6. Any special measures necessary for the accomplishment of the particular mission.

RESPONSIBILITIES OF THE DISPATCHING (BRIEFING) OFFICER

It is strange that so important a mission as that of patrolling can and does become such a haphazardous affair. The forces sending out the patrol are often vague about the mission and careless in the selection of the patrol leader and patrol personnel.

The failure of patrolling missions is not always due to mechanical failure of the patrol during the actual operation. Patrol failure due to improper briefing and careless planning of patrol missions by the higher commander or dispatching officer is not infrequent.

Patrolling should not be a function that is ordered and undertaken on the spur of the moment. Ordinarily normal patrolling demands can be foreseen far enough in advance by the higher commander or the dispatching officer so that a definite, well-planned patrolling schedule can be established.

Patrolling, which is one of the most difficult of combat operations, can become one of the most mishandled if the dispatching officer fails to give the patrol time for proper preparation, all available information and the facilities to expedite and prepare for its mission. The dispatching officer (who acts for the higher commander) is responsible for the following:

(1) *Selection of the patrol leader.*

(2) *Giving a specific mission.* The patrol should have one principal mission. Whether or not other missions can be performed or other information secured, incident to the principal requirement, must be left to the discretion of the patrol leader. The patrol must not be overburdened

with multiple missions of equal importance. Written questions in order of importance which the patrol leader will be expected to answer on his return may be issued to him at the time of briefing.

A secondary mission may be given which the patrol will undertake if certain designated factors affect the primary one. Care must be taken to carefully delineate the primary and secondary missions. It must be specifically understood when and under what conditions a multiple mission is to be undertaken.

(3) *Designation of the general route to be followed and time alloted for accomplishment of the mission.* Specific routes should be left up to the discretion of the patrol leader who will decide upon them after his personal observation and study of the terrain to be covered. All information available from prior reconnaissance, maps, photos, and other sources should be made available to the patrol leader. A patrol must not be expected to accomplish a mission in a few hours that will take an entire day or night to perform. Time and space factors based on careful study and previous experiences must be considered in assignment of time for the duration of patrol missions. It is well to stress that enemy information must be back in time to be of operational value; however, careful prior planning of the time and space factor will normally answer this basic intelligence requirement.

(4) *Giving the patrol leader the situation (enemy and own troops).* It is advisable to let the patrol leader see the situation map of the parent unit and explain to him the dispositions and all other pertinent information concerning the enemy, his own, friendly and adjacent units. He must be informed of other patrol activity and missions being performed by his own and adjacent units. If more than one patrol is being sent out from the same unit all patrol leaders should be briefed at the same conference so that each has a general knowledge of the objectives, general routes and missions of the other patrols.

(5) *Notification of the outpost and other security elements of the patrol mission and time of return.* Outguards and all other forward elements through which the patrol must pass and return (own and adjacent units) must be notified. Specific points of departure should be designated and provisions made whereby the patrol leader can receive from these forward elements the latest enemy information relative to his mission. The correct challenge password and reply must be given the patrol leaders and they should be instructed to check their accuracy as they pass through the forward elements. Provisions for guides to lead patrols through our own minefields and wire must be made.

(6) *Designating size, method of selection of patrol members, weapons and equipment (special and otherwise) of the patrol.* In deciding actual make-up and equipment of the patrol, the patrol leader can be consulted in regard to these details. This is an advisable procedure particularly if the patrol leader is an experienced one.

(7) *Giving all other pertinent information and special instructions,* such as:

a. Designation of special individuals to accompany the patrol (engineers, counter-intelligence men, or language specialists).

b. Designation of any portending change in the location of the CP of the unit which might occur while the patrol is still out.

c. How, where and when messages are to be sent back.

d. Information as to impending action which might affect the patrol mission, such as artillery barrages or air bombardment.

e. Providing and expediting the procurement of equipment and information necessary for the mission.

f. Insuring that the patrol has proper rest and food, prior to, and after the mission.

g. Action to be taken when the patrol returns. Where, when, and to whom reports are to be submitted.

(8) *Provide the means by which proper study of the terrain to be crossed can be conducted.* This can be done by conducting the patrol leader to an OP for his study and by the use of aerial photos. Terrain briefing by members of the photo-interpreter teams attached to divisions and regiments is advisable as they can present a detailed, recent and accurate study of most terrain conditions and obstacles. Aerial photos, as well as maps, can be carried on the mission by the patrol leader. Also, sand tables have been successfully used by the briefing officer.

(9) Especially for longer range patrols, primarily in the jungles, the possibility must be considered that the patrol will be cut off completely. In this case a point should be designated which, at a future date, will be occupied by our forces, and the patrol be told to proceed there to reestablish contact.

PATROL INTERROGATION AND REPORTS

Proper interrogation of returning patrols and accurate patrol reports are as important to the success of patrolling operations as proper briefing.

Upon the return of the patrol from the mission the patrol leader should make an oral or written report to the dispatching officer (or other designated individual). In a moving situation an oral report usually suffices, but when time and other circumstances permit, it is advisable to make a written report. Maps and overlays should be used for reference and clarification of written reports. If possible, interrogation can best be conducted in view of the terrain covered.

Often a patrol returns from a mission and turns in a negative report on its activity, when in reality it picked up information of value incidental to its principal mission which could be brought out by skillful questioning on the part of the interrogating officer.

Patrol leaders, patrol members, and messengers should all be questioned thoroughly upon return. Unknowingly, they may have observed information of value. Simplicity of mission is still a guiding factor in assignment of patrol tasks, but all information incidental to the principal requirement of the patrol must be obtained from the returning patrol members.

As in observation, a knowledge of the reliability of the individual patrol member being questioned is desirable, so that a proper evaluation can be made of his answers.

Men who have just returned from a patrol mission should be questioned immediately, but due consideration must be given to their physical and mental state, which is often affected by the strain of the patrol operation.

Patrol interrogation can also be done by the use of prepared blank forms containing questions which can be filled out by all members. The interrogator can also use leading questions to draw out many required facts. A map or aerial photo should be used during interrogation to help facilitate questions by the interrogating officer and to orient the patrol member being questioned.

At times men will remember things of importance only after they have had a chance to become rested and the tension occurring during the mission has died down. Questions pertaining to the principal mission must be asked immediately; information incidental to the main mission can be obtained at this same time, but it is often better to wait, if circumstances permit, until the patrol is rested.

As in the interrogation of prisoners of war, it is sometimes desirable to question patrol members separately or to have them fill out individual report forms. This procedure is particularly necessary when on a static front: Small but important details of change (which may have been observed but not mentioned in the first report) can be developed.

Inexperienced patrol members should always be closely interrogated because many times they will not appreciate the significance of things they may see on their mission.

RESPONSIBILITIES OF THE PATROL LEADER: As soon as the patrol leader receives his orders he selects the members (if they have not already been designated) and assembles the individual patrol members or unit. Alternate leaders and specific duties of the individual members will be assigned. The situation and mission will be covered thoroughly. Uniforms, weapons, and other equipment will be prescribed.

Circumstances permitting, he will designate periods for rehearsal and any additional briefing felt necessary. A time for reassembly will be set, keeping in mind the time necessary to perform rehearsals, inspections, and other final, detailed briefing.

Between the preliminary briefing and the time for reassembly the patrol leader must study all available maps or photos, and if possible, make a physical survey of the terrain to be covered. This will usually be done from an OP. His second in command and as many more members of the patrol as is feasible should be taken on this preliminary observation of the terrain to be traversed. During the survey and study, the terrain, formations, rendezvous points, and the specific route will be decided upon. Prominent landmarks to help maintain direction and danger areas will be noted.

If the dispatching officer has not contacted the PI team, and one is available, he should make arrangements for a detailed briefing on the terrain to be covered by one of its members.

Upon reassembly of the patrol, all points decided on during the terrain survey will be covered in detail by the leader. Rehearsals will then be conducted. (See Rehearsals). If time and circumstances permit, an enforced rest period should be designated after rehearsals, until time for final assembly prior to departure.

Inspection: As soon as the patrol assembles for departure, the patrol leader inspects it to determine the following:

(1) That each man is still physically fit for duty.

(2) That each man is in proper uniform and carries the equipment and armament prescribed. (Weapons should have been recently firetested.)

(3) That all identifying insignia and papers have been removed.

(4) That no equipment or clothing glistens or rattles.

Formation, duties, recognition signals, passwords, mission, and situation are then stated again in a final briefing. The patrol leader then reports to the dispatching officer that his patrol is ready for the mission (unless specified otherwise).

On leaving friendly lines, the patrol leader informs the nearest outguard or front-line unit of his proposed route and obtains any recent information they have concerning the enemy and friendly patrols operating in the vicinity. He may inform the front-line unit when and where he expects to return.

THE AMBUSH

The ambush makes use of the basic element of surprise, and is one of the most difficult forms of patrol operations. It is used for the destruction or capture of the enemy and it requires careful preparation, selection of appropriate terrain, and absolute team work by members of the ambush patrol. That an ambush can be successful without careful preparation and proper planning is the exception rather than the rule.

A knowledge of the enemy is essential. The size of his patrols, tactics, routes, formations, and regular habits, whatever they are, must be carefully considered. When these factors are known, an ambush, which ideally should be twice the enemy strength, can be formed.

The type of fixed ambush that will be used, either "silent" or "fire", should be decided upon in the planning stages and all terrain features that will aid in laying the desired type of ambush must be utilized. The type of ambush used will also influence the type of terrain selected. In reconnoitering the terrain to select a favorable site, the patrol leader should seek places of forced passage which an enemy patrol must use, such as bridges, cuts, passes, or other well-defined terrain irregularities. These terrain features can be selected initially by the use of the map, but if possible, they should be reconnoitered in person before an ambush is laid. Other terrain features, such as cliffs, swamps, water barriers, woods and other types of cover should also be considered in any plans for any ambush. A factor of considerable importance in such planning is the weather. Fog, rain, snow or other conditions producing poor visibility play an important part.

After the terrain has been selected, and the location and the duties of each member of the ambush patrol have been determined, a dress rehearsal should be held in areas in the rear and on terrain as similar as possible to the terrain chosen for the actual ambush. Friendly patrols can be used to simulate the actions of the enemy in these rehearsals.

The armament of the ambush patrol should be employed in relation to the terrain features involved. Normally, however, an ambush party will have light machineguns, issue sidearms, assault rifles, such as the M-16 and AR-15, and smoke and offensive grenades. There will be certain situations when mortars, heavy machine guns and antitank weapons can be used most effectively.

After the chosen ambush formation has been practiced and the rehearsal done to the point of proficiency, the ambush is laid.

If the enemy does not fall into the trap the first time, the selected point of ambush need not necessarily be changed nor abandoned. Even the most carefully chosen locations and well-laid plans are still subject to the actions of the enemy, and if the ambush does not succeed at the initial attempt, it is usually advisable to maintain it in the same area until there is little chance of success remaining. Never play hide and seek with the enemy. A constant changing in location of ambushes in order to intercept enemy patrols who did not run into the ambush the first time will result in an inadequate preparation and poor security as well as in counter-measures from enemy patrols.

The assault, whether "silent" or "fire" must be coordinated. *An easily recognizable signal for the assault must be used.* The action itself must be broken off as quickly as possible and a retirement completed before the enemy can cut off a possible retreat or fix the ambush party with fire. Sometimes an ambush party will have two groups rush an enemy patrol while the third group remains out to cover the action of withdrawal or reinforce the other groups, if needed.

A means of identification to differentiate between members of the ambush patrol and the enemy is a very important factor, especially at night when many ambushes will take place. To avoid possible confusion, white buttons, white handkerchiefs, luminous markers, or any similar kind of identification marker are recommended.

Fundamental requirements for successful ambushes are:

(1) The signal for assault must be clearly understood and easily distinguishable and obeyed implicitly by all members of the ambush party.

(2) Catch the enemy in terrain unfavorable to him. There must be full use of terrain and artificial obstacles, such as wire gaps, cuts, trails, and fences. Obstacles should exist between the enemy and ambush party when possible. Controlled smoke is of great use in ambush to create confusion and to help divert enemy fire. It also has an important use by a patrol that is ambushed, to allow escape.

(3) There must be a recognition signal for members of the ambushing patrol.

(4) Tracks, debris, and all other give-away indications should be eliminated around the scene of an ambush. The position should be occupied secretly and the area should be searched thoroughly for enemy counter-ambushes prior to occupation, and then occupied from the rear.

(5) Thorough knowledge of terrain, the enemy, and a simple plan based upon a clear and unmistakable signal, is absolutely necessary.

(6) Coolness is mandatory, especially from the first detachment to perceive the enemy if any ambush is to work successfully.

(7) Place men in effective groups so that each will be sure to see whether a patrol is enemy or friendly.

(8) Coordination achieved through rehearsals or the use of experienced men is vital to the success of an ambush.

(9) A rendezvous point definitely known to all must be designated prior to the action.

(10) The leader of a large ambush party should not actively commit himself to the fire fight if he can avoid it. He should remain free so he can see and control the action.

(11) A reserve should always be held out, especially if the ambush is a large one, and other enemy units are operating in the immediate vicinity.

(12) Special snipers should be designated to pick off officers and other key men at the time the fire fight begins. Snipers may also be placed so as to pick off any survivors of the initial assault.

THE FIRE AMBUSH

In a fire type ambush weapons are coordinated with terrain. In this method a fire line parallel to the direction of approach of the enemy is laid. Automatic weapons are placed at each flank. (Other light automatic weapons can be used to build up the fire line if available.)

This line will be laid taking full advantage of terrain features, such as defiles, cliffs, streams or artificial obstacles such as bridges, gaps in wire or fences. Any place of forced passage, where the enemy patrol will be canalized and cover and concealment is available for the ambushing party, can be utilized. The main body of the enemy patrol should be allowed to pass in front of the fire line before the assault is opened. Scouts and points should be allowed to pass the main elements of the ambush. They can be picked off by men detailed for this purpose. Smoke and mortar fire can be used to prevent withdrawals or canalize the enemy action into other prepared fire lines.

The reserve element can be used to cut off enemy escape or to cover the withdrawal of the ambushing party if enemy reinforcements arrive.

Double fire lines (on each side of the enemy) can be used if one line is above or not subject to fire from the other. Grenades and demolitions may be utilized particularly if the fire line is on high ground.

THE SILENT AMBUSH

The silent type of ambush which is often used for the taking of prisoners can only be successful and "silent" if the patrol to be ambushed is small and its strength is known. It is usually used at night.

A formation is laid taking advantage of cover and terrain where each member of the ambushing party is responsible for a particular member of the enemy patrol. (This is usually designated by enemy positions in file.) All members of the patrol remain perfectly silent and the enemy is allowed to penetrate until each man of the group is covered. On a given signal a simultaneous assault is made. Again the signal must be very clear and definite.

It must be recognized that silent ambushes have their limitations and that silence and secrecy are the main considerations. A patrol maintaining proper formations and interval will be difficult to ambush in this manner.

If the capture of prisoners is the main objective and silence after the silent assault and capture of a few men is not necessary, the ambush can be laid accordingly. Firearms can then be utilized against enemy personnel who cannot be attached physically.

Often times the silent ambush is not for the capture of prisoners but for the security reasons wherein the main purpose will be to kill the enemy silently, preventing enemy knowledge of the presence of the patrol. In this situation the decision whether to attack or rely upon cover and concealment is up to the patrol leader.

It is a common conception that in case of a silent ambush the enemy must be outnumbered two to one. The reason being that each enemy soldier will be attacked from both sides by two members of the ambushing party. This procedure will work, but confusion during the assault, coupled with poor light and terrain conditions may result in casualties being inflicted by members of the ambushing party upon each other.

Silent ambushes are made increasingly difficult because enemy patrol members will often carry grenades and automatic weapons ready for instant use, particularly in areas where cover is thick and they may be subject to close quarter attack. In this case as in all others, the members of the ambushing party must be in cover extremely close to the enemy. If they are not, enemy weapons may be brought into play and silence will be broken. Silent ambushes will usually be most successful when used in close proximity to enemy lines, where enemy soldiers feel more secure and are less alert.

The taking of prisoners silently is best accomplished by small patrols who will operate near enemy positions and pick off isolated sentinels or other outguard elements.

The *ambush of opportunity* is unlike the two fixed types of ambush wherein the time for preparation and planning is present.

Patrols often unexpectedly encounter the enemy in situations where they will lack the time or opportunity to lay the ambush so as to get the maximum advantage. When a meeting engagement is only foreseen in enough time to allow the patrol to take available cover and positions (so as to take advantage of the element of surprise), the results of the ambush will depend entirely upon the control, experience and training of the patrol members. In this type of engagement the experienced patrol will win.

At times the patrol may accidently come upon a similar enemy formation in which the element of surprise will be equal on both sides. Here again the patrol with the best training and experience will react more swiftly and come out on top.

COUNTER MEASURES AGAINST ENEMY AMBUSH

A well-trained, properly rehearsed patrol is the best counter measure against enemy ambush. If its leader properly employs a good scout (point) to thoroughly explore all enemy ambush capabilities before the patrol is committed, most enemy ambushes can be avoided or made less effective. The enemy ambush will ordinarily not get the entire patrol if proper intervals are maintained.

The same analysis used to lay an ambush should be considered by the scout or patrol leader in order to reveal the enemy's ambush capabilities. When a small, lightly armed patrol is ambushed, the best thing to do is to disperse. In such a situation, each man should proceed separately to the rendezvous point (which should have been predetermined by the patrol).

In the case of a small patrol, it is usually inadvisable to fight it out with an enemy ambush on the latter's chosen terrain. If the patrol is a large, well-armed one with a support element, decisive action may nullify the initial advantage of surprise by the enemy. The ambushing party can be wiped out or forced to withdraw.

PRECEPTS OF SUCCESSFUL PATROLLING

"If the mission is important give the patrol all possible help. If it isn't important, don't send a patrol."

J.M.

Modern combat has evolved the following precepts and techniques used by successful patrols. Combat lessons learned from the mistakes of soldiers now dead and techniques used successfully by our own troops and the enemy are listed here for use as they may apply to the particular situation and patrol:

(1) Since the advance of a patrol has much in common with the stalking of the hunter, if at all possible, it is advisable to advance down wind from the enemy. This is vital because the wind, no matter how weak, clearly carries all the noise that passes through it. With a medium wind prevailing, a patrol of fifteen men can be heard at three or four hundred yards even though all other factors are in its favor. Whenever possible, a patrol or scout should operate with the wind blowing in his face.

(2) At night the patrol should always try to move on low ground with the moon behind. Experience has prove that a man's view is more accurate and distinct when his back is to the moon. When looking toward the moon, the view is distorted by a type of haze or aureole.

A night patrol can see objects more clearly when they are silhouetted against the sky. For this reason a patrol or scout should operate at a lower elevation than the enemy. This is the reason patrols utilize the bottoms of vales and cross slopes and ridges with extreme care. This is particularly important in sectors where terrain features are very distinct. In such areas where friendly and enemy patrolling is constant, the possibility of contact between patrols is always present.

Colors are not distinguishable at night, since the eye can only distinguish between light and dark. Objects at night have no perspective, as surfaces only (not depth) can be perceived. For example, a hedge row seen from the side or obliquely, may look like a single bush or a thicket and the interior corners of a building are blotted out. Objects also appear larger and more distant at night than in the daylight.

Two of the best times to start patrol actions are at dawn and sunset. By using these tactics, the rising or setting sun is in the enemy's eyes, making it harder for him to see you. Observation at night can usually be carried out more effectively from a prone position. The eye should not be strained by concentrating on one object too long. If objects blur, the eyelids should be lowered slowly, kept closed for a few seconds, opened slowly and the eyes refocused.

(3) Delay can be avoided by moving boldly when firing is heard. Push forward when such sounds as shelling, wind rustling, or local fireing mask the sound of movement.

(4) Sounds of men walking or vehicles moving are heard more distinctly if an ear is placed close to the ground. Sounds are also transmitted farther in wet than in dry weather. Extremely cold weather also aids in the transmission of sound. Sounds are audible at greater distances on quiet nights than in the daytime.

(5) A threatened sneeze may often be stopped by pressing upward with the fingers against the nostrils. A threatened cough may often be stopped by a slight pressure on the Adam's apple.

(6) Do not smoke or use lights on a night patrol. If lights must be used, centralize such use in the hands of the patrol leader. Luminous dials on compasses and watches are visible to enemy eyes at close range. Keep them covered. Smoking during patrol halts should be kept to a minimum. Aside from the sight of smoke and sign left by discarded butts, the odor of the smoke will cling for hours, particularly in a damp area. All these things may betray the patrol to the enemy.

(7) When a man hears the sound of a flare leaving its discharger, he should drop to the ground before the burst. If he is caught unexpectedly by a bursting flare, he should freeze in position and remain motionless until the light dies down. The best time to move is just after the light has gone out. Care must be taken not to be caught in motion by recurrent flares. When illuminated by a flare, the eyes should be lowered until the light goes out so as to avoid temporary blindness. To freeze when caught by an enemy flare is usually correct but to simulate a tree in an open field that has been under observation by the enemy all day is foolish. When there is much flare activity on the part of the enemy, it usually indicates that he has few patrols out.

(8) Even unidentified persons in supposedly friendly areas who are met while on a patrol must be challenged. One member of the patrol should make the challenge from a concealed position. The other members will cover him. At night, the challenging soldier should take a prone position on the ground so he can silhouette the individual in question against the sky. Every sign of movement in "no-man's land" should be treated as evidence of the enemy until otherwise proved. When a patrol is returning from a mission and must pass through a friendly outpost line, one member should advance and make the contact. The main body of the patrol should wait until mutual recognition is established.

(9) Patrols must be equally careful in returning from a mission as in starting out, so as to avoid hostile patrols. Many patrols become too careless near their own lines; they forget that the enemy also has patrols out. Patrol formations and habits should be changed often to deceive the enemy and make his counter patrol action more difficult.

(10) Patrol action is often best performed during inclement weather because it affords concealment and finds the enemy less alert. Another good time to patrol is when the enemy is eating, especially if he has been discovered to have regular eating hours or habits.

(11) A speaking knowledge and ability to recognize a few key enemy words and phrases may be useful to the patrol member or scout. Words, such as "halt", "friend", "who's there", "advance", "fire", "take cover", etcetera, have been used to advantage in certain situations. A study of the use of these enemy words should be made. In most languages a few words and phrases can be learned and pronounced well enough to deceive the enemy. A man who speaks the enemy language can often be most useful near the point of the patrol, especially at night.

(12) A knowledge of the footprints of the enemy and his tire and vehicle tracks should be acquired as soon as possible. A knowledge of your own tracks and vehicle

markings should be acquired during the training period.

Cover all tracks when possible. Cigarette butts, paper, footprints, empty cartridge cases, bruised and broken vegetation will tell a story to the enemy scout.

(13) Native guides are very useful in reconnaissance. Allied troops who know the area should also be used as sources of enemy information. Anyone knowing the country or people should be used as an aid where security permits. Natives in uncivilized areas often have uncanny ability at reading signs, detecting ambushes, and maintaining direction.

(14) Grenades, both offensive and smoke, are very useful on patrol action. Certain patrol members can be designated to carry them in their hands for immediate use.

(15) When taking prisoners (especially at night) make them come to you. Don't relax. Make them stretch out their fingers when they raise their hands as they may have grenades concealed in them. Have the bulk of the patrol concealed, covering the men accepting the surrender. At times, surrendering enemy soldiers may fall on their faces when acting as bait by surrendering, permitting their concealed comrades to open fire over their heads.

(16) On day and night patrols, especially night, it is advisable, when there are several changes in direction necessitating different azimuths, to pre-set a compass for each change in direction prior to leaving on the mission. These compasses can be carried by individual members of the patrol who can be numbered off. For instance, when the second change of azimuth comes about the number two man will pass his compass up to the point man for use. The azimuth already being set. This is a very advisable method of procedure, because lack of time, light or cover may make it difficult to allow a reshooting of a compass azimuth while on the patrol mission.

(17) If a patrol can be armed with the enemy's small automatic weapons when making a penetration of enemy outposts, the enemy can often be confused by receiving return fire from his own weapons, which he is able to recognize by sound and flash. The individual patrol can often take advantage of the resultant confusion.

Caution must be exercised when enemy weapons are carried, because friendly patrols and units may mistake the patrol for the actual enemy because of the use of enemy weapons. Use of enemy weapons on patrol action will depend on the local situation.

(18) Patrols operating to the front can often be helped by prearranged harassing (artillery, mortar, MG) fire upon another area. Such fire will help divert enemy attention and allow the patrol to advance. In close-contact operations it is advisable for the patrol leader to check with the artillery liaison during the briefing period for knowledge of contemplated artillery barrages that might interfere with the patrol action. Arrangements can also be made for defensive or supporting fire to help the patrol in an emergency. Mortar fire has been used in some situations preceding the patrol as it advances forward on a night mission.

(19) A patrol, before going on a mission, must have plenty of rest. Men who have not had enough sleep and rest, physically and mentally, prior to the mission will go about it half-heartedly, and poor results will be obtained. Rest is also essential when the patrol returns from its mission, and every effort should be made on the part of the dispatching officer to see that the patrol is well fed, has the best available quarters and that the individual members have time to relax and recover from the physical and mental strain which they have undergone on the mission.

(20) Benzedrine has been used successfully in patrol work. It helps to keep the members alert, awake, and provides the necessary physical and nervous energy for a period as long as 6 hours. Normally a tablet is taken one hour after the patrol departs on the mission. Its effects will usually be felt about 2 hours after taking it. The energy provided will last for an additional 6 hours. More can be taken if the patrol is out for a longer period of time. Such a drug should be used in special situations and must not be relied on consistently to get good patrol performances. A quantity of morphine in syrettes or tablet form should also be taken along by the patrol.

If a member of the patrol becomes a casualty, and he cannot be brought back with the patrol, he should be given morphine to sustain him until medical aid can reach him. The lives of other members of the patrol should not be risked in an effort to bring back a casualty who impedes the progress of the patrol and prevents proper fulfillment of the mission. In a battle, casualties will always be left behind, except in case of special circumstances, these eventualities should be discussed and outlined prior to the patrol mission. Extra men can be taken on large patrols to bring back casualties. Medical aid men may also be taken along.

A definite plan to care for any wounded should always be made. Aside from the physical aspects of taking care of the wounded, the morale effect on the individual patrol member, who knows he will not be abandoned if a casualty, is very great.

Liquor as a stimulant should not be used prior to departure on a mission. If available, it can be issued just before an action, when the short lift (45 min.) it gives will be of benefit. On return from a mission liquor may be issued for its morale effect and to relieve tension while interrogating the patrol members.

(21) In moving over sharp rocks or thorny areas use the bear crawl to keep the knees off the ground. By using a clenched fist in thorny areas, unpleasant minor wounds on the inside of the hand can be avoided. The bear crawl, which is a rapid means of crossing areas where the cover is three feet high or better, should also be used at night when speed is paramount. The fewer parts of the body touch the ground, the more speed can be developed. Adhesive tape on elbows and knees and gloves will often

help if a great amount of crawling is anticipated by the scout or patrol member.

(22) When a patrol unexpectedly encounters a similar enemy patrol at close quarters, the patrol members must open fire immediately. It is especially important that all patrol members be adept in the use of all basic weapons. Combat firing (instinctive pointing) as well as aimed fire should be covered in training.

(23) Occasions will arise when enemy out-guards, or sentries, must be dispatched silently. Special training in sentry killing and stalking must be given. Patrols equipped with silent weapons must have had practice and training in their use. Results will be better if certain naturally qualified men are given such training and used accordingly.

(24) Sneak patrol members must be careful of the use of firearms while on operations. Promiscuous use of these weapons is liable to draw enemy fire. The small sneak patrol lacks sufficient strength to become involved in a fire fight. It should fire its weapons to only avoid capture.

(25) If animal or bird calls or sounds are to be used in signalling or in maintenance of control, practice in the use of these sounds must be undertaken. Unnatural imitations of nature are worse than no signals at all. In some situations bird calls are inadvisable for signalling as terrain that is much fought over will not contain birds.

Control at night can be facilitated by using luminous buttons or markers on the backs of the patrol members. On some occasions a compass tied on the back with the luminous dial exposed has been used. A compass can also be used by the patrol leader for signalling. By holding it open in the hand the luminous dial will show distinctive motions of the arm and hand signal. (Care must be taken when in close contact so that the luminous markers do not expose the patrol to the enemy.)

(26) In spite of thorough preparation a patrol may find itself lost. As many cross checks as possible should be used to prevent this. Position of the moon, direction of prevailing wind, compass bearings, or permanent objects, should be utilized. When lost the patrol leader should establish an all around defense and then move out from his patrol in a north, south, east and west direction until he has oriented himself. The patrol should not be dragged around with the leader while this is taking place.

(27) When grenades are thrown at night on any patrol action, the patrol members by closing their eyes before the flash will avoid temporary night blindness.

(28) In static situations it is just as important to know the location of friendly mine fields as those of the enemy. This is especially important when new units have replaced those in the line.

(29) The length of the mission (distance) as well as the size and type of patrol will often determine the type of armament. In other words short patrol missions such as encountered on a static front can be undertaken with the members carrying lots of automatic fire power. Long missions will require a decision on armament as automatic weapons and ammunition are heavy and tiring to carry.

(30) Snipers with telescope sighted rifles are often valuable on patrol missions. These men can cover forward areas while the patrol advances, cover withdrawals and can also be used to pick off enemy snipers, officers, machine gunners, or protect a withdrawal.

(31) Gum chewing, or tobacco chewing, while on mission should be permissible. Tension or nervous strain is relieved by such practices. Naturally, in the case of tobacco, the distinctive noise of spitting should be considered. Gum chewing by patrol members has also been found to cut down the tendency of some men to cough.

(32) If whispering is necessary for control or communication within the patrol it is best to expel the air from the lungs prior to the actual whispering. This avoids the very audible hissing sound which is often uncontrollable, particularly when the individual is under strain.

(33) If travel along a gravel road is necessary it is best to walk in the wheel tracks where the gravel is packed down. The shoulders of the road may offer silent walking conditions, but they should always be considered as likely spots for mines.

(34) If friendly and enemy wire are to be encountered on the mission, rehearsals and necessary training measures should be undertaken so that all patrol members can successfully pass this obstacle.

(35) In daytime patrolling it is often impossible to move forward along a completely covered route. If any choice is available and ground devoid of cover must be crossed, choose the route with the poorest cover at the beginning so as to utilize natural terrain features and cover when close to the enemy.

(36) When an area subject to enemy automatic fire must be crossed, a knowledge of the type and ammuition capacity of the enemy weapon is useful. The actual crossing being made while the enemy is reloading.

(37) The scout or patrol member must develop the habit of stopping frequently while on the mission so as to look and listen. This procedure should be S.O.P. During each stop the scout should "freeze". By "freezing" during the halt, energy is saved and the possibility of movement disclosing the position to the enemy is lessened. "Freezing" does not mean absolute rigidity such as standing at attention. There should be nothing strained about it, and all of the individual's power should be concentrated on seeing, hearing and thinking with the body relaxed and immobile.

The closer the contact with the enemy the more frequent the halts. You cannot hear well enough while moving. If you make a noise such as breaking a twig, you should "freeze" for a long period before proceeding. The enemy may be at rest and have you at a disadvantage. On such occasions it becomes a battle of patience. Get in the habit of putting yourself in the enemy's place. Never underestimate him. When you are moving assume that he is listening. Give the enemy credit for common sense.

Judge him as your equal and act accordingly because reconnaissance in close proximity to the enemy is often a battle of wits. When hiding in close proximity to the enemy, the eyelids should be kept almost closed. The eyeball will reflect light under certain conditions revealing an observer to the enemy.

(38) On a very dark night or in very dense cover it is often difficult to maintain direction. When you stop, do not alter the direction of your feet unless you can fix upon some object in line with your previous movement. Searchlights and tracers fired into the sky have been used successfully to orient night patrols who have been forced to operate without landmarks or other means of orientation.

(39) Be sure to synchronize watches among all members of the patrol. The patrol time should be the same as that used by the dispatching unit.

(40) The use of trained dogs to accompany patrols has proven valuable especially in detecting an enemy ambush. If the enemy is using dogs to guard his installations or on patrols, the difficulty of the mission will be greatly increased. Surprise and close contact reconnaissance are very difficult to achieve when dogs are employed by the enemy.

(41) Utilize all possible means of communication, control and recognition. The portable radio is useful on short sneak patrol missions and can be used for inter-patrol communication, such as between forward and rear elements of large patrols. The new portable lightweight radios are especially useful for contact with the parent unit. All possible utilizations of sound powered phones, signal lamps, colored lenses on flashlights, colored smoke and flares should be considered when planning a patrol mission. Communication check points should be established along the route beforehand and at a designated time contact should be established from these points with the parent unit. In this way the dispatching officer can follow the patrol's progress.

(42) If a "get-away" man is appointed, his reliability must be proven, as there is a tendency to flee at the first sign of action and bring back alarming and premature reports of disaster to the patrol.

On long range missions assignment of progressive rendezvous points and proper planning will usually make the use of the "get-away" man unnecessary.

The radio may be used by the patrol to direct artillery fire on enemy targets of opportunity. This can often be done without interfering with the principal patrol mission.

(43) Have all the men take care of the necessary urination and bowel movements prior to departure on the mission. This reduces subsequent delays and does away with another possibility which may endanger the patrol.

(44) Although patrol security, by use of flankers and points, is mandatory, there is a danger that the main strength and source of fire power of the larger patrol can be wasted away by having too many men on the points and flanks. Keep the main body of the patrol where it can be controlled and its fire power utilized. Don't disperse the patrol strength too much.

(45) When messages are designated to be sent back at specific times during the mission, consideration must be given to the effect of this procedure on the operational strength of the patrol. Extra men must be added for this purpose.

When an unforeseen incident of sufficient importance occurs that must be transmitted to the unit, any patrol member may be dispatched regardless of the effect on the operations of the patrol.

Messengers returning from patrols should always be regarded as sources of recent information. Aside from the message they bring, they should be interrogated by the recipient of the message.

(46) In thickly wooded areas where isolated snipers are likely to be encountered it is often advisable to employ a double point for the patrol. The two men at the point operating closely together will often discourage the sniper from firing, as he can only fire at one at a time, thus revealing his position to the other.

(47) The best men should be used on the point and flanks of a patrol but it must be remembered that the tension under which they operate and the extra terrain they must investigate to protect the main body will cause them to tire faster. The patrol leader must make the necessary reliefs at the points and flanks to prevent this from becoming excessive. As men tire, they become less attentive and careless.

(48) It is dangerous to split up a patrol once the line of departure has been crossed. If this procedure is necessary and could not have been forseen, a very clear system of recognition, definite redezvous point and mutual knowledge of each group's mission must be present.

(49) Silent movement is often dependent upon silent footwear. If issue shoes are not sufficient to meet the demands of silence, sacking, portions of inner tubes, or other materials can be wrapped around the shoes to help muffle noise.

(50) In extremely difficult terrain where darkness and dense foliage make control difficult, a cord connecting (single file) members of a patrol can be used for control. These cords can be used to maintain direction and to signal (by jerks) the various patrol members. Naturally, rough terrain will render this means of control and signalling more difficult, but this system has been used with good effect in some situations. The morale effect of the patrol members' knowing they are in contact with each other by means of the cord is great.

(51) Captured vehicles and parts of enemy uniforms may be used to achieve surprise, deceive or in baiting of ambushes.

Some sneak patrols, spotted by the enemy, have set up dummies and retired while attention was distracted. Similarly, some patrol members have hung jackets on wire and placed helmets with the tops showing to hold enemy attention.

Figure 64. A six man diamond formation. Notice zones of observation of each man. The leader moves about within the formation to maintain control. Note: the distance between men has been reduced for illustration purposes.

If an enemy sentry must be by-passed or killed, divert his attention by throwing stones, breaking sticks in the opposite direction, enabling him to be approached or by-passed. Small patrols can also be used to distract the enemy's attention by noise and fire so that a larger body can reach its objective.

(52) Patrols operating in enemy rear areas can create great confusion by turning sign posts and direction arrows around. At times this procedure has been successful in directing the enemy into prepared ambushes. Telephone wire cut at night with one end drawn into a prepared ambush spot will often lure repairing linemen to it, so they can be killed or captured.

(53) A patrol approaching a populated locality of unknown sympathies can often draw enemy fire forcing him to disclose his presence, by breaking out into an open place on the edge of the locality and firing in the general direction of the village, then taking cover. The enemy if present will often open fire.

(54) At times a scout or patrol member may be forced by circumstances to feign death to escape detection. This can best be done by placing himself as close as possible to other dead soldiers (own or enemy) and remaining there until the danger is past. A helmet or jacket with a bullet hole in it will help add realism.

(55) A member of your patrol or ambush party may be wounded and fall away from cover. If he calls for help, do not go to his aid immediately, as the enemy will cover the space around the wounded man with fire. Likewise, a wounded friendly soldier has often been used as bait for an ambush.

(56) Patrols operating at night have successfully designated enemy positions to artillery observers by the following means:

a. Use of flash lights to draw fire.

b. Placing of flares (ground-like used on RR) near positions where artillery fire was desired.

c. Using tracer fire in the sky to form an X above the enemy position.

(57) Patrols have been ambushed or driven into areas of prepared fire and demolitions by the following stratagem. Light automatic weapons have been placed in trees along trails, etcetera, so that they can be operated from a concealed position by means of a wire or string attached to the trigger. Upon receiving fire from an unexpected and seemingly innocent area, the patrol members have instinctively dispersed into prepared areas where the enemy was waiting.

Figure 65. A three man patrol formation. Notice how all around observation is maintained.

Figure 66. A four man diamond patrol formation. Notice how each man observes in his assigned direction. Note: the distance between men has been reduced for illustration purposes.

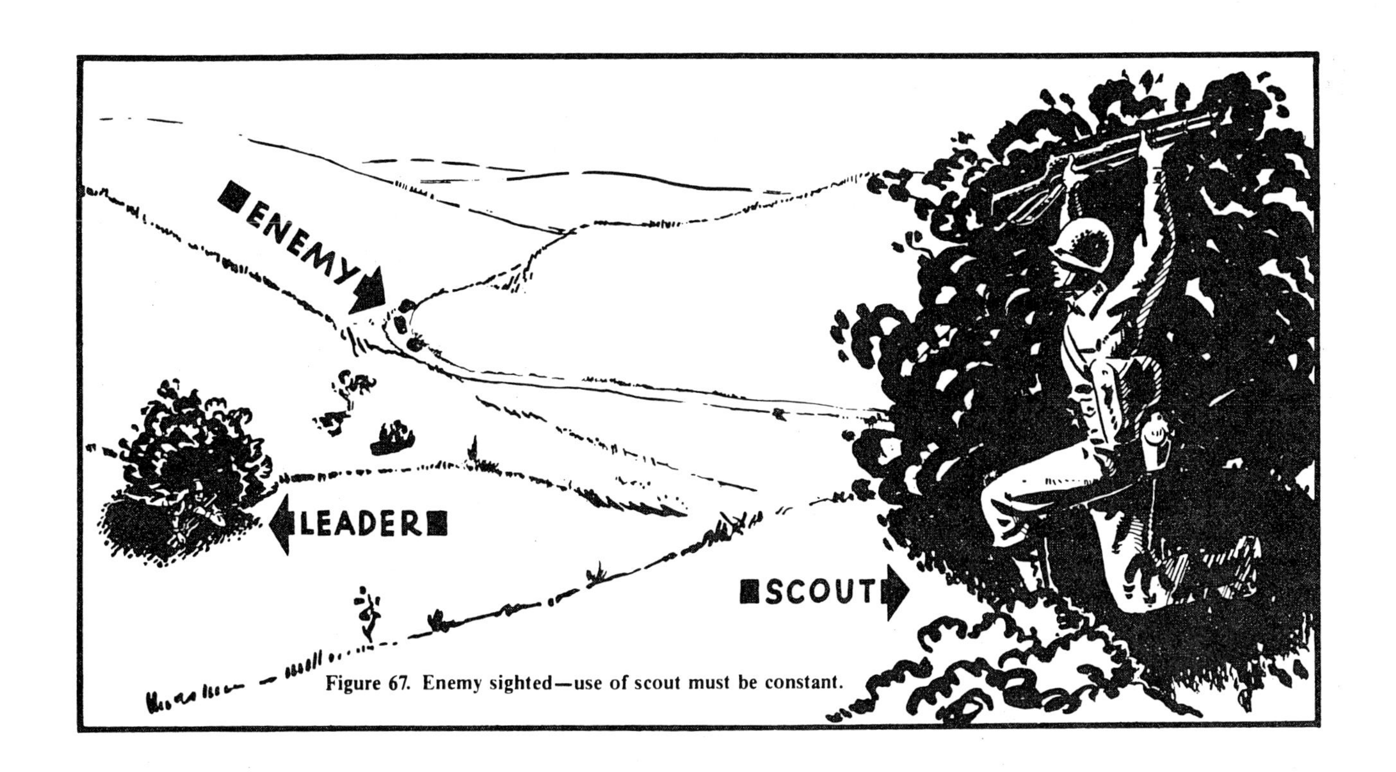

Figure 67. Enemy sighted—use of scout must be constant.

Figure 68. In daytime, utilize cover.

ROGERS' RANGERS

Tactical principles in warfare remain unchanged. History can always draw a parallel to any tactic used in modern battle. Modern warfare differs from that of the past principally because it utilizes newer and deadlier weapons. Present day terminology employed in description of tactics will differ from those used by Caesar but the fundamentals of tactics and the use of terrain were as well understood by the great generals of history as they are by our present day generals.

Like large scale tactics, patrolling techniques and methods of the present day have their counterpart in history.

In the period 1760-1765, Robert Rogers organized a band of backwoodsmen into a fighting unit that was later to become famous as Rogers' Rangers. The present day American Rangers have taken their name from this famous unit, since immortalized in the fine book and motion picture, "Northwest Passage".

The Rangers operated with success on the side of the British in the French and Indian War, being principally used for missions of raiding and combat reconnaissance. Individually each Ranger was mobile and self-sufficient. He killed quietly, took prisoners (if any) with stealth and speed, and was a past master in the use of cover and concealment. His courage and hardihood made him feared and famous.

The "Regulations for Rogers' Rangers", which follow, are so similar to present day patrol tactics that explanations and comparisons are almost unnecessary. Except for the difference in equipment and terminology, these regulations can be adapted almost in their entirety for present day patrolling.

Figure 69. Security at the halt—must provide you with all-around protection.

Lieutenant General Vandegrift of Guadalcanal fame best expressed these thoughts by the following statement:

"My message to the troops in training is to go back to the tactics of the French and Indian War. Study their tactics, fit in our modern weapons, and you have a solution. I refer to the tactics and leadership of the days of Rogers' Rangers."

I. All Rangers are to be subject to the rules and articles of war; to appear at roll-call every evening on their own parade, equipped, each with a firelock, sixty rounds of powder and ball, and a hatchet, at which time an officer from each company is to inspect the same, to see they are in order, so as to be ready on an emergency to march at a minute's warning; and before they are dismissed, the necessary guards are to be draughted, and scouts for the next day appointed.

II. Whenever you are ordered to the enemies' forts or frontiers for discoveries, if your number be small, march in a single file, keeping at such a distance from each other as to prevent one shot from killing two men, sending one man or more forward, and the like on each side, at a distance of twenty yards from the main body, if the ground you march over will admit of it, to give the signal to the officer of the approach of the enemy, and of their number, etcetera.

III. If you march over marshes or soft ground, change your position, and march abreast of each other to prevent the enemy from tracking you (as they would do, if you marched in a single file) till you get over such ground, and then resume your former order, and march till it is quite dark before you encamp, which do, if possible, on a piece of ground that may afford your sentries the advantage of seeing or hearing the enemy some considerable distance. Keeping one half of your whole party awake alternately through the night.

IV. Some time before you come to the place you would reconnoiter, make a stand and send one or two men to whom you can confide, to look out the best ground for making your observation.

V. If you have the good fortune to take any prisoners, keep them separate, till they are examined, and in your return take a different route from that in which you went out, that you may the better discover any party in your rear, and have an opportunity, if their strength be superior to yours, to alter your course, or disperse, as the circumstances may require.

VI. If you march in a large body of three or four hundred, with a design to attack the enemy, divide your party into three columns, each headed by a proper officer, and let those columns march in single files, the columns to the right and left keeping at twenty yards distance or more from that of the centre, if the ground will admit, and let proper guards be kept in the front and rear, and suitable flanking parties at a due distance as before directed, with orders to halt on all eminences, to take a view of the surrounding ground, to prevent your being ambushed, and to notify the approach or retreat of the enemy, that proper dispositions may be made for attacking, defending, etcetera. And if the enemy approach on your front on level ground, form a front of your three columns or main body with the advanced guard, keeping out your flanking parties, as if you were marching under the command of trusty officers, to prevent the enemy from pressing hard on either of your wings, or surrounding you, which is the usual method of the savages, if their number will admit of it, and be careful likewise to support and strengthen your rear guard.

VII. If you are obliged to receive the enemy's fire, fall, or squat down, till it is over, then rise and discharge at them. If their main body is equal to yours, be careful to support and strengthen your flanking parties to make them equal to theirs, that if possible you may repulse them to their main body, in which case push upon them with the greatest resolution with equal force in each flank and in the centre, observing to keep at due distance from each other, and advance from tree to tree, with one half of the party before the other ten or twelve yards. If the enemy push upon you, let your front fire and fall down, and then let your rear advance through them and do the like, by which time those who before were in front will be ready to discharge again, and repeat the same alternately, as the occasion shall require; by this means you will keep up such a constant fire, that the enemy will not be able easily to break your order, or gain your ground.

VIII. If you oblige the enemy to retreat, be careful in your pursuit of them, to keep out your flanking parties, and prevent them from gaining eminences, or rising grounds, in which case they would perhaps be able to rally and repulse you in their turn.

IX. If you are obliged to retreat, let the front of your whole party fire and fall back, till the rear hath done the same, making for the best ground you can; by this means you will oblige the enemy to pursue you, if they do it at all, in the face of a constant fire.

X. If the enemy is so superior that you are in danger of being surrounded by them, let the whole body disperse, and every one take a different road to the place of rendezvous appointed for that evening, which must every morning be altered and the whole party, or as many of them as possible, together, after any separation that may happen in the day; but if you should happen to be actually surrounded, form yourselves into a square, or if in the woods, a circle is best, and if possible make a stand till the darkness of the night favors your escape.

XI. If your rear is attacked, the main body and flankers must face about to the right and left, as occasion shall require, and form themselves to oppose the enemy as before directed; and the same method must be observed, if attacked in either of your flanks, by which means you will always make a rear of one of your flank-guards.

XII. If you determine to rally after a retreat, in order to make a fresh stand against the enemy, by all means endeavour to do it on the most rising ground you come at, which will give you greatly the advantage in the point of

situation, and enable you to repulse superior numbers.
XIII. In general, when pushed upon by the enemy, reserve your fire till they approach very near, which will then put them into the greatest surprise and consternation, and give you an opportunity of rushing upon them with your hatchets and cutlasses to the better advantage.
XIV. When you encamp at night, fix your sentries in such a manner not to be relieved from the main body till morning, profound secrecy and silence being often of the last importance in these cases. Each sentry therefore should consist of six men, two of those whom must be constantly alert, and when relieved by their fellows, it should be done without noise; and in case those on duty see or hear anything, which alarms them, they are not to speak, but one of them is silently to retreat, and acquaint the commanding officer thereof, that proper dispositions may be made; and all occasional sentries should be fixed in like manner.
XV. At the first dawn of day, wake your whole detachment; that being the time when the savages choose to fall upon their enemies, you should by all means be in readiness to receive them.
XVI. If the enemy should be discovered by your detachments in the morning, and their numbers are superior to yours, and a victory doubtful, you should not attack them till the evening, as then they will not know your numbers, and if you are repulsed, your retreat will be favoured by the darkness of the night.
XVII. When you stop for refreshment, choose some spring or rivulet if you can, and dispose your party so as not to be surprised, posting proper guards and sentries at a due distance, and let a small party waylay the path you came in, lest the enemy should be pursuing.
XVIII. If in your return, you have to cross rivers, avoid the usual fords as much as possible, lest the enemy should have discovered, and be there expecting you.
XIX. If you have to pass by lakes, keep at some distance from the edge of the water, lest, in case of an ambuscade, or an attack from the enemy, when in that situation, your retreat should be cut off.
XX. If the enemy pursue your rear, take a circle till you come to your own tracks, and there form an ambush to receive them, and give them the first fire.
XXI. When you return from a scout, and come near our forts, avoid the usual roads, and avenues thereto, lest the enemy should have headed you, and lay in ambush to receive you, when almost exhausted by fatigue.
XXII. When you pursue any party that has been near our forts or encampments, follow not directly in their tracks, lest they should be discovered by their rear-guards, who, at such a time, would be most alert; but endeavour, by a different route, to head and meet them in some narrow pass, or lay in ambush to receive them when and where they least expect it.
XXIII. If you are to embark in canoes, battoes or otherwise, by water, choose the evening for the time of your embarkation, as you will then have the whole night before you, to pass undiscovered by any parties of the enemy, on hills or other places, which command a prospect of the lake or river you are upon.
XXIV. In paddling or rowing, give orders that the boat or canoe next the stern most, wait for her, and the third for the second, and the fourth for the third, and so on, to prevent separation, and that you may be ready to assist each other in any emergency.
XXV. Appoint one man in each boat to look out for fires, on the adjacent shores, from the numbers and size of which you may form some judgment of the number that kindled them, and whether you are able to attack them or not.
XXVI. If you find the enemy encamped near the banks of a river or lake, which you imagine they will attempt to cross for their security upon being attacked leave a detachment of your party on the opposite shore to receive them, while, with the remainder, you surprise them, having them between you and the lake or river.
XXVII. If you cannot satisfy yourself as to the enemy's number and strength from their fire, etcetera, conceal your boats at some distance, and ascertain their number by a reconoitering party, when they embark or march, in the morning, marking the course they steer, etcetera, when you may pursue, ambush and attack them, or let them pass, as prudence shall direct you. In general, however, that you may not be discovered by the enemy on the lakes and rivers at a great distance, it is safest to lay by, with your boats and party concealed all day, without noise or show, and to pursue your intended course by night; and whether you go by land or water give out parole and countersigns, in order to know one another in the dark, and likewise appoint a station for every man to repair to, in case of any accident that may separate you.

Such in general are the rules to be observed in the ranging service; there are, however, a thousand occurrences and circumstances which may happen, that will make it necessary, in some measure to depart from them, and to put other arts and strategems in practice; in which cases every man's reason and judgement must be his guide, according to the particular situation and nature of things; and that he may do this to advantage, he should keep in mind a maxim never to be departed from by a commander, viz: to preserve a firmness and presence of mind on every occasion.

Extract from: "The Bulletin of the Fort Ticonderoga Museum", Volume VI, January 1941, Number 1.

(See suggested lecture on patrolling using Rogers' Rangers as its theme.)

WORLD WAR II
FOREIGN GROUND RECONNAISSANCE

The following section is excerpted from Col. Applegate's original manual that was used to train special Allied troops during WW II. It is included here because knowledge of foreign reconnaissance methods is helpful when designing similar training programs. For historical purposes, it is presented in its original form, and in the present tense.

Both author and publisher hope that the reader will appreciate and use these discussions, for they helped establish the basic rules of reconnaissance that international armed forces still use today.

BRITISH

The British emphasize thorough specialized training of officers and men in scouting and patrolling. Many are graduates of the Commando Training Schools. Their patrols are usually led by officers and the basis of their specialized training rests on the recognized principle: "THE IMPORTANCE OF OBTAINING INFORMATION SUPERIOR TO THAT OF THE ENEMY CANNOT BE OVEREMPHASIZED."

Before reconnaissance begins the British consider two aspects: (a) is the enemy stationary? (b) moving? In the first instance, having evaded, outwitted, or brushed aside the opposing reconnaissance elements, probing takes the form of combat reconnaissance patrol thrusts across the whole extent of each patrol's zone of reconnaissance, until a clear picture of the enemy's outpost zone has been obtained. In the latter instance, probing is executed as permitted by the moving situation—usually from a reconnaissance base.

The English believe that it is the duty of the scout and the observer to supply their commanders with information which cannot usually be provided by ordinary infantry troops. These specialists must be able to obtain accurate information under all conditions of warfare, in all kinds of terrain. Responsibilities placed on their reconnaissance personnel are:

(1) Small and large scale patrolling, observing in small detachments, sniping and oral reporting.

(2) Penetrating enemy lines and working inside them.

(3) Moving skillfully and silently over difficult country at night.

(4) Constructing field defenses and erecting obstacles.

(5) Executing demolition and sabotage.

They insist that their scouts and observers be physically fit and one hundred per cent self-confident and that their men be competent at interpreting aerial photographs, executing field sketches, performing first aid, riding a horse, driving a motorcycle, sailing a boat, swimming and cooking.

Good scouts, the British insist, should always have the answers to the following questions:

(1) Are enemy patrols on foot, motor, or mounted?

(2) What is their average strength? How armed? How many vehicles?

(3) Do they patrol in armored cars or tanks?

(4) At what times do they use the routes?

(5) How will they summon assistance when attacked?

(6) From what point will they arrive?

(7) What kind of supply and troops? Trained or untrained? Fresh or tired?

(8) What kind of terrain?

The British insist upon careful planning of not only patrols but aiso of ambushes. They say that in choosing the position of the ambush, it is necessary to take the following requirements into account:

1. A safe, sure line of retreat such as wooded or broken ground.

2. Firing positions from which fire can be opened at point-blank range.

3. At least two firing positions on opposite sides of the road.

4. If possible, the locality should permit the ambushers to see the enemy while he is from three to four hundred yards away, so that if his strength is dangerously superior, he may be allowed to pass.

The British realize that all ambushing parties, allied or enemy, usually assume similar set-ups. Consequently, they insist that reconnaissance patrols be instructed to study the ground not only from the tactical point of view, but also to the possibilities of suitable camouflage. The British also point out to their ambush patrol leaders that many dead spaces exist within enemy positions and that the average patrol does not investigate these dead spaces skillfully. For this reason, such a dead space would be the ideal position for ambush.

The most fitting climax for any discussion of combat and reconnaissance patrols is found in a letter from a young British officer to a friend in which he describes the tactics of the Moroccan Goums:

"The best patrolling troops we have come across are the Moroccan Goums, whose success as compared with any European unit, is phenomenal. Even against the best Germans, they never fail. Why are they better than we? First, because they are wild hillmen who have been trained as warriors from birth. Second, the preparation of their patrol is done with such detailed thoroughness. No fighting patrol is sent out until its leaders have spent at least a day watching the actual position they are after, reconnoitering exact routes and so forth. If the leaders are not satisfied at the end of the day, they will postpone sending out the patrol, and will devote another day to preliminaries."

The British list the following as "Golden Rules" for patrol work:

1. Know your mission. Understand it specifically and ask questions about it until you do.

2. Study your terrain. Make a thorough study of the terrain over which you are to go and keep studying it as you advance.

3. Make your plans brief, simple and specific.

4. Stop at the outguards; ask them for the latest information and give them some when you return.

5. Make regular reports.

6. Observe carefully.

7. If necessary, use fire reconnaissance and force the enemy to reply.

8. Support the advance of one patrol by the fire of the second whenever possible.

9. Avoid fire fights, when possible.

10. GET THE INFORMATION ** GET IT BACK ** AND GET IT BACK IN TIME!

RUSSIAN RECONNAISSANCE

The Russians, like the Germans and the Japanese, have placed great emphasis upon active ground reconnaissance as the principal means of obtaining tactical information. Russian reconnaissance generally is extremely aggressive, and reconnaissance in force is a standard Russian tactic. The importance placed on active ground reconnaissance is also indicated by the size of their tactical reconnaissance unit, although only limited information is available.

The Army is divided into basic units, such as infantry, artillery, engineer, cavalry, armored and air, all of whose fighting elements have their own reconnaissance agencies. The strength of divisional reconnaissance agencies—the infantry—is as follows: each battalion has a reconnaissance platoon, each regiment has a reconnaissance company, and each division has a reconnaissance battalion.

As far as known, a motorized infantry division has a reconnaissance battlion composed of:

1 armored car company (3 platoons).

1 artillery battery (four 76.2 mm guns).

1 motorized infantry company (120 men, 6 MG's).

It is reported that the infantry division has in its reconnaissance battalion the following elements:

1 battery artillery (four 76.2 mm guns).

1 tank company of 3 platoons (16 T38's).

1 AA battery (eight 4 barrel MG's).

1 cavalry troop (126 horses and men).

1 motorized infantry company (10 cars, 120 men).

1 armored car platoon (four 75 mm AT guns).

This battalion executes all special reconnaissance missions and has the necessary numerical and armored strength to make use of reconnaissance by force tactics. In addition, it can be used for special patrols in raiding, harassing or other fighting missions. It should be noted that in almost all cases the strength of the Russian reconnaissance units is three times greater than that of comparable American units.

The reconnaissance company of the infantry regiment (I & R Platoon in the American Army) is commanded by the regimental intelligence officer who is usually a major. This is another indication of the importance the Russians place upon their ground reconnaissance units.

In addition to the organic reconnaissance agencies, partisan (guerrilla) units which are trained for operation behind enemy lines furnish information. In a retrograde movement the members of this partisan force who are principally armed with grenades, submachine guns and demolitions, remain in concealment while the enemy advances. Such units are supplied by air and are used for harassing, raiding and reconnaissance. Their information of the enemy is delivered to the main Russian forces by radio, air or individual scouts filtering through from behind the enemy lines. The use of partisan companies has proved of great benefit to the Russian command who visualized their possibilities and trained them especially for this work prior to the war with Germany.

The Russians are the greatest believers in specialized training of troops of any of the warring powers. They conceive of an individual soldier's or unit's being trained in all combat phases but organically complete units are made of experts in one particular field. They "tailor make" reconnaissance units to fit the mission by drawing from these specialized units. The specially trained scout-observer is the backbone of all Russian reconnaissance.

All Russian soldiers are trained in reconnaissance and observation, but in every infantry squad there is one soldier who has been taken from the unit and given extra special training in scouting, observation and sniping. This soldier is called a scout-observer. Highly trained men of this type working in pairs or in small groups, perform most of the necessary observation and sneak patrol missions. The Russians maintain that the scout-observer must be trained in enemy tactics, armament, and recognition, that he must have good physical and moral qualities, be in good health, courageous, have good hearing and eyesight, and be endowed with patience and tenacity. He is also trained in terrain appreciation, mapping, and sketching.

These scouts are the principal source of intelligence and operate ceaselessly once contact with the enemy has been made. They are sent out to cover enemy front lines and are constantly seeking the location of limiting points and flanks of the enemy's battle formations. They are alert for any movements, sound, or muzzle blast which disclose position of pill boxes or installations. When the enemy is in retreat, these scout-observers maintain close contact and strive especially to locate the flanks and points where the enemy is attempting to reorganize. They are often assigned to kill enemy officer personnel and are highly trained in the sniping phase of rifle marksmanship. Often these scouts are organized into special "destroyer" units detailed to wipe out patrols who try to infiltrate Russian lines. On combat reconnaissance patrols of special importance the entire group may be made up of scout-observers. In almost all cases regardless of size or composition of larger patrols, scout-observers perform essential duties at the point and flanks.

ITALIAN

The best information on Italian reconnaissance principles comes from a manual issued by the Italian army in 1942. Their principles are as sound and complete as those of any army, yet combat reports indicate that the Italians, as a whole, did not apply the principles laid down in their manual with any great degree of success.

Although they do use units and individuals from regular infantry companies for scouting and patrolling, they do not believe it is a sound principle to rely on them for all reconnaissance purposes. The infantry battlion had a reconnaissance platoon consisting of one officer and forty-four men. This platoon and its individual members were not to be used for any other purpose than that for which the men had been trained, scouting and patrolling being their only combat functions. The members of the platoon were to be picked from the outstanding men of the various companies in the battalion, the selection being made by the company commander. The men selected were to have the following qualifications: courage, level-headedness, initiative, physical capacity, military knowledge, skill in weapons, and intelligence.

Once this selected platoon was formed, it was to be directly responsible to the battalion commander, and he alone was responsible for its training and employment, their being no S-2 as we know it in the Italian army organization. The training of the individual members of the platoon included the following subjects:

1. Orientation—cardinal points, compass, terrain feature, etcetera.
2. Range estimation.
3. Perception of signs indicating presence of the enemy.
4. Ability to determine size and type of enemy unit (by depth, road space, etcetera.
5. Hand to hand combat training, marksmanship.
6. Interrogation of prisoners and civilians.
7. Knowledge of most common means of communication.
8. Clearness and precision in preparing messages.
9. Knowledge of terrain features and their proper names.

The patrol was to be trained in the following:

1. Rapid movement.
2. Proper use of terrain for cover and concealment.
3. Appropriate action in encounter with enemy units and in case of ambush.
4. Movement over varying terrain and under lifferent degrees of visibility.
5. Approach to and assault of small enemy units using only bayonets and knives (with a minimum of noise).
6. Camouflage.
7. Searching of buildings.

The scout platoon, as it was normally called, was to have the following missions: reconnaissance, terrain intelligence, and the manning of observation posts. In performing these missions units or patrols of varying sizes were to be employed, the size being determined by the specific task.

THE RECONNAISSANCE PATROL

The reconnaissance patrol consists of a reconnaissance squad (or an ordinary rifle squad from the rifle company). Its mission is to obtain enemy information. A patrol may be assigned a large area to cover generally or a smaller one to reconnoiter thoroughly (the mission may be general or specific). The patrol leader is responsible for the performance of the patrol and the accomplishment of the mission. He must know beforehand exactly what information is wanted. The patrol is assigned a mission, an objective (or series of objectives), and a direction. On a general mission the following questions are asked: Where is the enemy? How many? How armed? What is he doing? Where is he going? On a specific mission the patrol answers the following questions: Is the enemy there? What type unit? Disposition: What is he doing? What are his outposts and other security units? An objective is usually also a rallying point. The direction is indicated by a road, a stream, or a series of easily recognized terrain features along which the route of the patrol lies.

All arrangements for sending back information are made beforehand: to whom the message is sent, how often, and how sent. The patrol leader should have a compass, binoculars, a notebook and a map or sketch of the area he is to patrol. The map or sketch should show features that are easily recognized on the ground, or that are otherwise important. The patrol leader checks his men and their equipment, and orients them (mission, route, etcetera).

The patrol moves from one observation point to the next along its route (on flat terrain, houses, steeples and trees may be used as observation posts). All cover and concealment is to be utilized along the route. On reaching an observation point, the patrol is halted under cover or in defilade and all necessary security measures are taken. Silhouettes against the skyline must be avoided. In crossing over a crest, all available concealment is to be used.

To reconnoiter a larger area along the route, second pairs of scouts are sent out. They avoid areas.where they enemy has been seen, going around them, if possible.

To examine minor terrain features along the route a pair of scouts is sent out and told where and how to rejoin the patrol. (This type of procedure is different than that of other armies.)

A patrol will be told before it leaves whether it is to fight or not. If it is possible to eliminate a small enemy unit quietly, it may do so; if not, the size and strength is determined and the patrol proceeds on its mission. In enemy-occupied areas the patrol will proceed on its mission by infiltration; if this is not possible, the patrol will remain in observation and send word to the C.O. In general the patrol will fight only to avoid capture or to return to its own lines. If the patrol is ambushed, or if it is necessary to warn the main body of the enemy's presence, the patrol may engage in a fire fight. In terrain not occupied by the enemy, the patrol returns by a different route. At night, in fog, or in wooded areas, however, the same route is used in returning.

SOP for Scouts

1. Have notebook and pencil.
2. Work in pairs.
3. Always act as though you are being observed by the enemy.
4. Except when necessary, avoid walking along the center of roads.
5. Cover open ground between objectives rapidly.
6. Having reached an objective, halt, observe to front and flanks, study the next objective and the route to it. Signal anything unusual to leader, and then continue.
7. Keep off skylines and away from prominent terrain features.
8. A house or other building is first observed from all sides and then entered by one scout while the other covers him from outside.
9. Signal the presence of the enemy to the patrol leader and act only on his orders.
10. In unknown territory orient yourself by the use of reference points.
11. Learn to depend on your hearing when visibility is poor.

Besides the normal reconnaissance patrol and the larger combat patrol, the Italians have two other types. The first is called a Terrain Intelligence patrol. It consists of a group of three or four men who are specially qualified, usually having been trained in engineering and surveying, and well equipped to study terrain features. This type of patrol is assigned the task of furnishing specific terrain information on assigned areas. Such things as man-made objects, water lines, bridges and fortifications are its objectives.

The other type is the Vigilance patrol, consisting of from five to twelve men. This patrol is nothing more than a security detachment, and its main mission is the operation of listening posts at night. It will be located at points of probable enemy approach, and will remain in position only during hours of darkness. It is always instructed to return to its own lines prior to daylight. The patrol members are instructed to remain alert with bayonets fixed. Small enemy units are allowed to approach and are captured by surprise without firing.

GERMAN RECONNAISSANCE

In the German Army battle and terrain reconnaissance is started when contact with the enemy has been made and is conducted as long as operations last.

In the standard infantry division, the German 1-C, equivalent to G-2 in the American Army, has a reconnaissance battalion of about 575 men. Its function is to hit, hold, and to keep moving forward.

Within the regiment, intelligence functions are performed by the commander or adjutant. He is assisted by two non-commissioned officers with language qualifications and a platoon of mounted infantry, which is actually part of the regimental headquarters company. It is composed of 32 horsemen, divided into three squads of eight men each, which may go out as four-man patrols, permitting six patrols to be in operation simultaneously. These patrols are frequently attached to battalions.

In the battalion, 1-C work is usually handled by the adjutant. His agencies are two: patrols attached from the regimental mounted infantry platoon, and patrols drawn from subordinate units. Within each company headquarters are two or three men specifically trained for scouting, patrolling and observation. The leader of the company headquarters section is responsible for "observation of the field of battle and scouting".

Since the Germans actually use combat to seek information regarding the enemy, reconnaissance units may also be used for other combat purposes. Before battle has been joined, the divisional reconnaissance battalion operates out in front looking for intelligence information, but when the battle begins, it reverts to the control of 1A, whose duties correspond to those of American G-3. Under his control, it may operate as a heavily armed combat battalion and 1-C gathers most of his information from battalions or regiments in the line. Besides his staff functions as adviser to division commander, the German 1-C also is the commander of his reconnaissance troops.

Usually the German reconnaissance battalion operates in three echelons. The first echelon is mobile and acts as a feeler; the second is mobile, more heavily armed, and acts as a support; the last is less mobile and very heavily armed to permit it to act as a combat support for the unit. Organization into three echelons is similar in armored, motorized, mountain, or infantry divisions. In the case of the infantry division, the forward echelon is a company of horsemen and the supporting element a company of bicyclists, both of which are very mobile. The general support echelon for the entire battalion is a company of heavy weapons which includes armored cars, engineer troops, anti-tank weapons, infantry howitzers, and heavy machine guns.

Within the German company are two methods of patrolling. The first, with reconnaissance as its only goal, is the observation patrol, usually consisting of three men, one or two of whom are officers, and a member of the trained observer personnel. One or both of the officers may be replaced by a non-commissioned officer, although usually the officer is used.

When reconnaissance patrols move out, the men are usually armed with small arms only. They reconnoiter until they find an area which warrants investigation, and then report its location to a foot combat reconnaissance patrol which performs the actual reconnaissance.

It is a common German tactic to use the combat patrol as a reconnaissance group to confirm observation or obtain information by fire reconnaissance.

The mission of the German combat patrol is to act as a reconnaissance screen, raid small units, harass the enemy, capture personnel, and disrupt lines of communications. The Germans, like the Americans, insist that every unit must be ready at all times to send out all types of patrols with varying strength and objectives. Careful preparation and study of latest enemy tactics are basic.

Although larger scale patrol activities, especially those involving enemy positions, are dependent on previous approval from division, constant and regular patrol activity is always maintained, even though the situation may be quiet or special orders for patrols have not been issued. This is not wholly in agreement with the American doctrine in which the battalion commander usually directs and restricts his subordinate leaders and gives specific orders as to when patrolling shall be carried on.

The following are prerequisites for all German patrols

1. Knowledge of objectives.
2. Knowledge of routes.
3. Knowledge of terrain.
4. Agreement on signal communication.
5. Complete equipment.

Before considering the size of the patrol or its techniques, it is important to examine the German ideas on patrol equipment. It is invariably necessary, the Germans say, to include at least one light MG in all combat or combat reconnaissance patrols. On many patrols the light mortar is also carried. In all cases where there is even the slightest possibility of encountering enemy armored vehicles, at least one anti-tank rifle and numerous mines should be added. In addition to these three essentials, many patrol members will carry the German machine pistol (submachine gun). The exception to the use of automatic weapons on German patrols is the mounted infantry, which, as was previously pointed out, carries rifles and has a MG section.

A formation often used by German patrols is the "T" or cross formation, a variation of a column formation, which has proven its value in all cases (say German documents) as the basic formation of advancing patrols. The T formation, of course, has to adapt itself to the ground: in open country and with a large patrol, the "T" will be relatively large; in more dense country, or if the patrol is small, the "T" will be smaller. At various times it will be necessary to detail connecting files to the center of the column; at other times the "T" may have to adapt itself to rough country by assuming an irregular form.

The usual strength of a German combat reconnaissance patrol seems to be between 16 and 20 men and an officer. Contrast this with the "T" formation diagrammed below, as used by American forces.

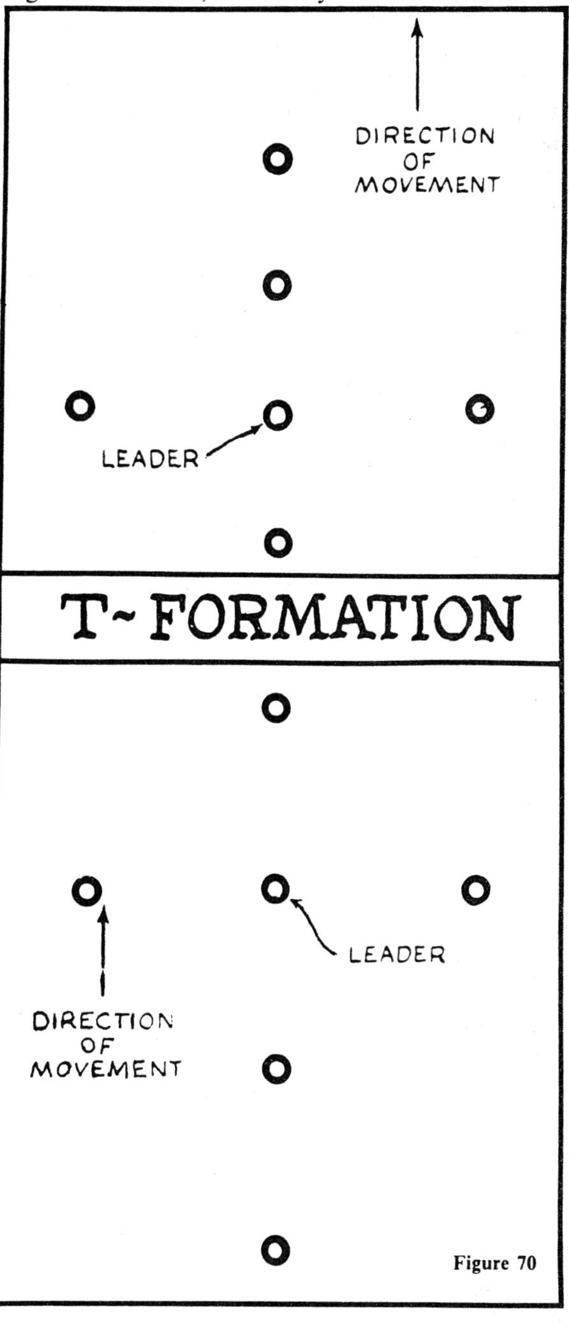

Figure 70

The German patrol leader is instructed that unnecessary casualties can be avoided by the following precautions:

1. The cross ("T") should never be smaller than is necessary to maintain visual communication. The cross, of course, varies with the terrain. For instance, in open flat terrain, they say that the cross should never be less than 150 by 100 yards.

2. The patrol should choose a different route of return both for simplicity and security.

If the patrol leader will remember the above two rules, and adapt the cross to the terrain, situation and mission, the Germans say that the danger of the patrol's being surrounded will be avoided. One of the principal missions of the German fighting patrol is to deny to the enemy the use of any observation posts which may have been established. The good patrol proceeds from one possible observation post to the next, making use of all roads and routes in the early stops, but later, as it approaches the enemy, making use of all available cover.

In movement in woods, formations in depth are used to facilitate control, mobility of leadership, and rapid transmission of orders. Dispersal of the patrol in the approach is, of course, regulated by the denseness of the woods. The Germans emphasize thorough reconnaissance of wooded areas and minute planning of the mission both of the large units and of the patrol. In all operations over wooded terrain, their troop units carry out continuous reconnaissance to avoid surprise attacks by the enemy. A typical German practice is to send several reconnaissance patrols abreast to the front and others to the flanks. In this connection, they are careful to make the distance between adjacent patrols wide enough for one patrol to avoid being confused by the noise of neighboring patrols; in thick underbrush this distance, they state, should be at least 160 yards.

Whenever possible, German penetration into wooded areas is accomplished by dividing the woods into sections to be cleared by patrols. The thicker the woods, the closer the formation. Before crossing any open space, thorough observation is made. Listening posts are established near the edge of the woods to keep open country under observation.

In line with the principle of silent movement, any equipment which might create noise or is cumbersome is left behind. The equipment and armament of the patrols usually includes machine pistols, rifles (with telescopic sights) and many egg grenades. The steel helmet is left behind. Each patrol carries at least two compasses, one in front and one in the rear, to guard against any deviation from the course.

The Germans are particularly adept at ruses to deceive enemy patrols, and at tricks which will enable them to penetrate enemy area with their own patrols. For instance, they frequently allow enemy sneak patrols to move unopposed through their positions, without opening fire, so that the location of their weapons will not be disclosed. If an enemy patrol should happen to stumble on a German position, the Germans try to kill or capture every man, so that no information of the position is reported.

Another trick of the Germans is to alter their dispositions after enemy patrols have withdrawn, a remedy for this type of action being to keep the territory under constant observation or seize and hold it until further elements can be brought up for the occupation.

All available information of the terrain is used to help the patrol which goes out from German lines. Since the object of the patrol is to obtain and deliver information about the enemy, it is especially necessary to obtain minute information about existing roads, paths, and clearings. Important conclusions concerning the future positions of the enemy can be drawn from the condition and location of tracks and paths. The location and condition of streams and ditches and bridges if of importance to their 1-C section, as well as to their 1-A.

Many other characteristics of the terrain are taken into account. Such points as the thickness of underbrush, height of trees, and the location of high or low ground and swamps are carefully considered in the patrol's observation, in addition to the usual military items.

Germans frequently try to draw fire on a large scale. They send out large night patrols to give the impression of an attack, and thus draw all types of fire and disclose the final protective lines. When the real attack comes, these positions are known, can be easily engaged, and the most dangerous areas avoided.

Whenever they suspect that new troops are facing them, the Germans send out a large number of patrols to locate and identify the units. Against inexperienced troops the Germans may use their infiltration patrols with great success. Any student of patrolling should understand the principle and tactics behind the employment of these patrols.

Automatic riflemen are recruited from the best and most experienced fighters in the German Army. At the beginning of the war, automatic riflemen captured at the Russian front stated that they enjoyed the privilege of longer furloughs after certain periods of service in the front line, and after the war they had been promised special subsidies and donations of land in Russia. In this way the Germans insured the greatest amount of initiative and best results on the part of their automatic rifle patrols.

The Germans have several ways of using their automatic riflemen, dependent on the nature of the engagement. During attacks, they are almost always used frontally in comparatively large groups of platoons or companies. In this case the goal is to trick the inexperienced commanders into premature employment of their main forces, a plan which was often successful during initial engagements in the summer of 1941. Inexperienced commanders often deployed the bulk of their forces to meet this front assault, which had every appearance of being made in vast numbers and with considerable firepower. In the meanwhile the German main columns (screened by the automatic riflemen and supported by artillery) endeavored to envelop and turn the flanks and strike in the rear.

In offensive combat the Germans seldom employ automatic rifle patrols in the early stages of the engagement. Acting cautiously, screened by reconnaissance and supported by units of the main forces, the German patrols sound out and probe the lines, trying to find soft spots—especially junctions between two defending units, or open flanks.

When such weak spots are found, they introduce automatic rifle patrols in groups of two or three. Carefully and skillfully using every terrain feature, woods, ravines, gullies and buildings, these groups infiltrate throughout the lines, mostly on the flanks, and penetrate the rear.

At the beginning of the engagement, they stay under cover. Sometimes they do not give themselves away for days, while they watch for developments; the idea being to join the combat only when it reaches a decisive stage. They select their firing positions and camouflage carefully. Gullies and ravines, deserted buildings, tree tops and shell holes are usual hiding places. They study the disposition of the opposing forces and watch every move and note every change in the situation. When the main forces join in combat, these hidden nests of fire suddenly come to life on the opponents' flanks and in the rear.

Materially, the results of these tactics are quite negligible. Their main mission is not to inflict casualties, but to create panic and confusion. Their fire is often unaimed and a great deal of ammunition is wasted in order to create a morale effect. Luminous bullets are often used for this purpose during the night. In the early stages of the war these tactics often succeeded, because green troops fell for this deception and numerous inexperienced commanders thought that they were encircled and in a hopeless position.

The success which has rewarded German reconnaissance is attributed to careful planning. All German patrols, whether of two or twenty men, seem to have been well planned and their personnel well trained. Maps are a basis for utilizing terrain in patrolling, but the German doctrine is that maps should be so thoroughly studied that they can be generally dispensed with during action. The map should be carried in the head, not the hand.

To the Germans, teamwork and preparation are the secrets of successful reconnaissance. They believe that haphazardly formed patrols made up of men who have never worked together are of little use. The Germans believe that best results are obtained from patrols which have been thoroughly trained under the most realistic conditions. In field exercises, the enemy is represented by actual soldiers who are dressed as the enemy and who fire and maneuver to approximate as closely as possible actual conditions.

Although the Germans use both small and large reconnaissance patrols, they find that their greatest success is gained from the combat reconnaissance type, or patrols formed from one or two squads and under the leadership of a platoon commander, supported by light machine guns and a mortar.

THE SQUAD IN SCOUTING AND OUTPOST DUTY

(extracted from a recent German Field Manual)

When used for reconnaissance and security for resting troops, the squad should be employed in its entirety whenever possible. Training in reconnaissance and security duties should be incorporated in the training of the individual as well as of the unit. The squad leader will have to be able to make decisions independently when on scouting or security duty. For this, he must be specially trained.

Strength and composition of the scouting party depend on situation and mission. Normally, the party will be of squad strength, 10 men, sufficient to neutralize enemy patrols or to by-pass them boldly.

In exceptional cases—missions depending solely on observation—a scouting party may be made up of a few observers only.

If enemy forces in defensive positions are to be reconnoitered, combat patrols in strength of several squads, led by the platoon leader and supported by heavy weapons, will force the enemy to show his weapons and his positions.

Each scouting party has a specific mission. The officer dispatching the party indicates clearly what he wants to know. Destination of the scouting party after completing the mission will be ordered.

The leader of the scouting party instructs his men as to the mission and the manner of execution planned. If every man in the party knows what is essential for executing the mission, success is guaranteed in case the squad leader is incapacitated.

The leader of the scouting party is to be equipped with a road map or sketch, field glasses, prismatic compass, watch, message pad, pencil, colored crayon, whistle and for night patrols, with a flashlight and a pistol.

Before departure, the dispatching officer may further order the following:

1. Special armament (sub-machine gun, AT rifle, smoke hand grenades) special ammunition or clothing (i.e., garrison cap).
2. Carrying of rations, full canteen.
3. Handing over of all written matter.
4. Arranging of password (at night).

When meeting other scouting patrols or security detachments, results of observations will be exchanged.

Following terrain features, the scouting party advances in bounds from observation point to observation point. The sectors in which the party moves will be increasingly smaller, the closer to the enemy the party advances.

When contact with the enemy becomes probable, the leader of the party will advance by stalking with part of the squad. The rest of the squad observes the stalking or follows ready to fire.

If scouting parties meet with enemy scouts or pickets, it may be to the advantage of the execution of the mission to avoid contact either through detouring or be letting the enemy patrol pass. If this is not possible, the enemy must be attacked in order to force reconnaissance of enemy main forces. If the scouting party meets with enemy scouts unexpectedly, it is essential to attack immediately.

Every opportunity to capture prisoners will be utilized. Prisoners are often the best means of obtaining enemy information. Scouts accomplish their mission only if reports come in promptly. The best report is useless if it arrives too late. Hesitation while scouting, and giving enemy patrols too wide a berth jeopardize the accomplishment of the mission.

The leader of the scouting party must decide whether and when to send messages, and must also determine how soon the message can reach the recipient. Unless ordered otherwise, first contact with the enemy will always be reported. A negative report (certain sector free from hostile forces) may be very important. The dispatching of messengers with unimportant reports decreases needlessly the strength of the party.

Each member of the scouting party observes the route of advance and tries to memorize prominent features in the terrain. In difficult terrain, during darkness and in thick weather, the road back (messenger route) must be marked in an inconspicuous manner.

If the enemy blocks the road back, another route must be chosen if possible. If impossible, attack.

Before an impending attack, scouting parties of sufficient armament to permit combat action will usually stay in contact with hostile forces and serve as security guards for the advance of their own units. They will later be absorbed without impeding fire.

During darkness scouts must often pause and listen or lie down and observe. If meeting enemy patrols unexpectedly, cut-and-thrust weapons are to be used primarily.

For purposes of reconnaissance or security permanent scouting parties may be advanced to suitable points of the terrain. They will stay there either until a predetermined time or until relieved.

JAPANESE RECONNAISSANCE METHODS

The Japanese Army operates on a theory of aggressive reconnaissance similar to German or Russian tactics. This type of reconnaissance is particularly suitable to the terrain and conditions of the Pacific theatre. The responsibility for all reconnaissance is placed in the hands of the commanding officer.

The reconnaissance agency for the Japanese Infantry Division is usually a reconnaissance battalion. The personnel of this battalion are infantry troops, with a total strength of about 450 officers and enlisted men.

The reconnaissance battalion is composed of four companies, and although there is no definite rule to be stated for their organization, one of the companies will almost always be an armored car company. The vehicles with which this company is equipped are fully tracked, making them actually tanks, known as "tankettes".

The other three companies of the battalion are either mounted or truck-borne infantry. Both the mounted and the truck-borne infantry companies are organized with a headquarters unit and two platoons. In the case of the mounted infantry, only infantry weapons such as light machine guns and rifles are carried. The truck-borne company, however, employs more fire power—one platoon of light machine guns and rifles, and a platoon of grenade dischargers or 37 mm infantry rapid fire guns. Men of these reconnaissance battalions are all basically well-trained infantrymen. Over a period of years of compulsory military training, the army has laid a sound foundation on basic and small unit tactics. This has made the Japanese particularly good at general reconnaissance and regardless of the branch of service, it can be assumed that the ordinary Japanese soldier, his noncoms and officers, are well-trained in squad and platoon tactics and patrol work.

In addition to the specialized Reconnaissance Battalion, the individual Japanese unit commander sends out his own patrols to cover the area to his immediate front. The numerous small infiltration patrols which penetrate enemy lines easily because of jungle growth and terrain are given great freedom of action, letting the patrol leader accomplish what he thinks he can do successfully with his patrol after his principal mission has been accomplished.

In considering the operations of Japanese reconnaissance agencies, it is important to bear in mind the enemy's idea on terrain. To him, foul weather or unfavorable beaches are not obstacles to reconnaissance; rather, they are conditions which are likely to result in a minimum number of our troops being on the alert or in the area, and therefore make movement safer and easier.

Also, the Japanese consider streams and rivers to be highways, rather than obstacles. In many instances, particularly in Malaya, their patrols made use of waterways for envelopment.

A large percentage of Nippon's army has been drawn from the peasant class. The men are used to living outdoors, and are familiar with the ground. They have never become familiar with those comforts and inventions of civilization which tend to spare the man, and consequently no great change in life is necessitated when they step into a uniform and go off to the hills and the islands and the jungles to fight. The Japanese, long knowing who and where he was going to fight, has been trained to the hardships of long campaigning under adverse conditions of weather and terrain. He is carefully trained in the art of natural camouflage. Elaborate helmets and body netting, and skin paint are also used.

Another common reconnaissance agency is the single scout. Many dead scouts have been found in observation posts around American strongpoints. At times, these scouts are unarmed, and only supplied with food; they relied for their protection on stealth and patience. They sneak in by night, observe during the day. These scouts are clever at concealment and take full advantage of trees, jungle growth, and other natural and artificial aids, and will remain motionless for long periods of time—an exhibition of patience which it is well to note. By heritage and rate the Japanese is patient. He can wait long hours for an action that may never come, and the important consideration is that he can wait silently and without movement.

The Japanese relied heavily upon individual riflemen (snipers) in the early stages of the war for harassing effects on the troops and for information. These riflemen were reported to be located in trees at the outset of the war, but lately have been concealed in individual fox-holes in the ground. They must be considered as a source of enemy information as well as in the light of other harassing effects upon the troops.

Japanese scouting and reconnaissance patrols operating in daylight consist of from 4 to 8 men, usually led by an NCO. They rely upon camouflage and stealthy movement and all are usually armed with a light machine gun and rifles, as well as plenty of grenades.

They move rapidly, in single file, employing no flank guard. This rapidity of movement precludes ambush in most case, since anyone seeing the patrol has little chance of moving ahead of them to prepare ambush. When close to the position to be reconnoitered, or when halted for any reason, the patrol establishes its own small perimeter of defense, and proceeds to observe and listen. Frequently one man is sent into a tree for better observation, and a member of the patrol invariably makes a hasty sketch of the terrain features in the vicinity and includes any information which may be gathered by the patrol while it lies in observation at that particular spot. Proficiency in sketching is good throughout the army; it has apparently been emphasized a great deal in the training.

Japanese patrols usually have the same type of objective as those of other armies. They strive to locate and learn the tactical ways of the enemy, inspect the terrain so as to be able to locate and sever the routes of supply which are so all important in the jungle. Patrols are also sent out on special missions such as capturing prisoners, blowing up ammunition dumps and of late have been used for counter-battery action against our artillery. The patrols dispatched to destroy our own artillery are specially trained in destroying these weapons and are usually sent out when the Japanese artillery is so weak that it cannot locate and destroy American guns by counter-battery fire.

The Japanese army has long recognized that the basic private cannot be trained to do everything in the way of special reconnaissance and patrolling missions. A policy whereby certain men from every infantry company are picked out and trained for additional specialty work such as counter-artillery action, destruction of ammunition dumps and other vital installations has been established. Consequently, when a commanding officer has a special patrol mission of this nature, he merely calls for so many of the specially trained infantry men from each unit to make up a patrol which is sent out on this particular mission.

Frequently, small patrols are sent out with their main mission to locate a position to which they can lead a combat patrol by night. These patrols usually leave their own area about noon, creep up on the objective, and lie in observation until dusk, when they return to lead the combat patrol.

The size of the Japanese combat patrol varies but is usually about eighteen or more men. These patrols strive to locate positions by firepower or act as harassing groups, attacking small isolated positions located by previous reconnaissance. They are usually equipped with LMG's, rifles, knives, plenty of grenades, and at least one grenade discharger. After dark they may split into two parts. One group, getting close to the enemy, commences calling and making other sounds to draw fire; the other group watches for muzzle boast, and movement, and creeps up on the flanks, attacking with bayonets, knives and grenades.

The Japanese often employ officer patrols to confirm previous information and such patrols are usually forerunners of an attack. This type patrol moves out late in the day, near sunset, to assure that important information picked up by the daylight reconnaissance patrols is still accurate, and to glean any other information possible, prior to launching the assault.

Another common Japanese patrol tactic is fire reconnaissance. One part of the patrol exposes itself, and draws the fire of our weapons. The rest of the patrol locates the positions from the firing and flanks it or returns immediately to report this information and lead a larger force to attack the disclosed positions.

The Japanese usually have scouting patrols ahead of the main body. Once they make contact with the enemy, they try to avoid frontal attacks. Instead, they send patrols around the flanks and to the rear of the opposition; these patrols are usually small, consisting of from two to twelve men. These men have apparently been trained and hardened for jungle warfare and they are given wide discretion as to tactics. The members of these patrols are usually expert swimmers, and have other qualifications for overcoming the difficulties of jungle warfare.

In order to make their movements through the dense jungle growth less obvious, the Japanese make trails no wider than is absolutely necessary. Their scouting parties may make trails by taking a long stick of soft wood, and with a circular motion, use it to push aside undergrowth. These trails are very difficult to detect and may easily be missed unless a marker is sighted. Such a marker may be a footprint placed on the side of the trail, or a small arrow pointing in the proper direction.

The Jungle Ambush

Japanese night attacks usually follow clearly defined terrain features, such as ridges, valleys, or crests. They are planned from information gathered by daylight observation. This is an important point as the Japanese invariably selects his objective for night attack from observation at sunset of enemy positions and dispositions.

Although it is not too difficult to ambush jungle trails, the most advisable procedure is to lay the ambush on the *main* trail, rather than a smaller secondary trail, of which there are many. The reason for this is that the complicated, intertwining system of small trails can turn an ambush into too much of a firefight.

A jungle ambush, like any other, relies for its success upon the element of surprise. And, more than in other situations, jungle ambushes should have a reserve to cope with the unexpected and to exploit favorable conditions and situations.

Japanese reaction to ambush has proved so similar in many cases that subsequent actions can be accurately predicted. The forward elements get off the trail, and attempt to move around the ambush. The rest of the patrol advances rapidly, and employing a mortar and LMG, attacks the ambush right along the trail. This action develops quite rapidly.

Japanese patrols have been successfully ambushed many times by placing "bait" on the trail. Such articles as steel helmets, discarded weapons, and so forth, are excellent because the Japanese have the tendency to cluster around them for an examination. Footmarks carelessly left on the trail by the ambushing party are quickly picked up by the enemy, and give the ambush away They can also be used to lay a "track trap" for ambush.

There are certain other characteristics and failings displayed by the Japanese which have contributed to the forming of a successful technique of laying an ambush against them. These characteristics make the Japanese particularly susceptible to ambushes.

Some of these characteristics are:

(1). They are often trail bound.

(2). Their security is inadequate.

(3). They tend to bunch up.

(4). Their "silence discipline" is poor; they jabber.

The Japanese react quickly and violently when surprised by an attack. Upon encountering a Japanese outpost and striking hard, destroying it or causing it to disperse, it is certain that Japanese combat patrols from adjacent sectors will move toward the point of encounter in order to reinforce the outpost under attack. Therefore, it is a good plan to ambush the approaches to the outpost in anticipation of this Japanese maneuver.

(1) The effect of an ambush depends on the surprise delivery of a maximum volume of close range fire, with the fire converging on the same target from at least two directions. Automatic weapons should be emplaced so that they will have converging fire at various points along the trail, from within the ambush.

(2) The ambushing force must be alert for any Japanese attempt to flank its position. The Japanese will almost invariably attempt such a maneuver. They employ both light and heavy mortars with speed and accuracy. A reserve should be held out to neutralize the Japanese mortars and destroy their crews, retaining the advantage gained from the ambush.

(3) The patrol must assign snipers the mission of shooting Japanese officers, non-commissioned officers, and Nambu (light machine) gunners.

(4) Routes of withdrawal from the ambush area must be selected. Rendezvous points must be designated, to which patrol members must go, if they are forced to disperse, or if they are separated. This will enable the patrol to reorganize and avoid confusion.

(5) The fire of the ambushing force will be held until the officer in charge gives the signal to commence firing.

(6) If the mission of the ambush is to inflict maximum casualties on the Japanese, then the leader will signify the accomplishment of the mission and give the signal for withdrawal.

(7) Ambush positions should be entered from the rear, because if the trail on the Japanese side of the ambush shows indication of American presence, the Japanese will be alerted. If the trail on the Japanese side must be used to enter the ambush position from the rear, all traces of American passage over the trail must be obliterated.

(8) If possible, anti-personnel mines or pits containing sharpened bamboo stakes may be hidden off the sides of the trail so that Japanese soldiers will be annihilated when they dive for cover.

(9) Japanese equipment such as helmets, rifles, gas masks, and the bodies of dead Japanese are excellent lures. The Japanese are very curious and will probably congregate around the body or the equipment.

(10) Maximum surprise is attained if a suitable position, not necessarily tactically ideal, is chosen for the ambush. The enemy might scout the tactically ideal position, assuming it to be the most likely ambush position, and discover the ambush before it can be executed. Observation and fields of fire can be sacrificed in order to gain maximum surprise.

(11) Care must be taken to place the various groups in a staggered formation if they are to be deployed on both sides of the trail. This will prevent the groups from firing into each other. A unit deployed on both sides of the trail should have its support group in a position most advantageous according to the terrain adjacent to the trail. The Japanese would probably have the choice between enveloping the group on one side of the trail, or that on the other. Mines could be planted with good effect off the side of the road opposite the ambushing unit, anticipating a Japanese move in that direction upon being surprised with fire from the ambush.

(12) If groups are sited on one side of the trail only, limits of fire must be established.

(13) In setting an ambush against the enemy, take time to find an innocent-appearing site, perhaps not ideally strong tactically, but one the character of which will not alert the Japanese on approach. They know a likely ambush site when they see one and are accordingly cautious.

(14) Emplace automatic weapons to cover the trail primarily, with secondary fields of fire on normal lines of deployment and withdrawal.

(15) Use high explosives (TNT, Bangalore Torpedoes) freely. Conceal them in the undergrowth adjacent to the trail and fire them electrically with the ten-cap hand exploder. High explosives so controlled and placed well down the trail will serve to pin in the Japanese patrol and prevent its escape.

(16) Study the approach from the enemy side and pick out those trees which appear likely to be selected for use by Japanese snipers. Work a few double-edged razor blades into the bark to discourage climbing.

(17) Be sure you have more than one reconnoitered route of withdrawal.

(18) Obliterate footprints made by the patrol in erecting an ambush. This is hard to do, but the technique can be picked up from natives.

COMBATTING JAPANESE AMBUSHES

The observation of strict trail discipline is the best means of combatting Japanese ambushes and robbing them of their effectiveness.

Factors in trail discipline are:

(1) Don't bunch up.
(2) Don't straggle.
(3) Don't talk.
(4) Don't lose contact.
(5) Don't relax vigilance.

The converging fire of Japanese automatic weapons is usually fixed, and often from two to four feet off the ground, making it possible to avoid the fire by crawling or creeping under it. An enemy ambush position must be enveloped swiftly. Speed will surprise and upset the enemy. Slowness will result in casualties.

Automatic fire should be held until targets appear that justify such fire. The ambushed force must bring fire on the enemy immediately. A definite plan of action should be decided upon by the patrol before leaving its base, assigning each element of the patrol a definite action in event of an enemy ambush from any direction.

Figure 71. Snipers are used to pick off leaders and key men.

CHAPTER V

LECTURES

The following four lectures are offered as samples of methods of presentation of fieldcraft and patrolling.

A SUGGESTED LECTURE ON FIELDCRAFT

Training Aids:

1. If the class is small, a sand table may be used for demonstration purposes.

2. If the class is large and if the facilities are available, use a film strip or glass slides in a darkened room.

3. As a third alternative, each student can be issued a map to follow or a large map may be drawn on a blackboard.

THE SITUATION:

Cpl. John Thomas is one of the surviving members of an ill-fated patrol. He has escaped an enemy ambush and at 1700, December 194__, he is standing on a ridge above Ricker Farm in enemy territory. (Sabillasville sheet of the Gettysburg Antietam series 1:21,120).

THE MISSION:

To return to a previously picked rendezvous point on the outskirts of Sabillasville.

TIME:

The patrol members had previously agreed to stay at the rendezvous point until 2100.

EQUIPMENT:

Cpl. Thomas has a map, a carbine, a compass and his wits.

PREPARATION:

Cpl. Thomas has already devoted considerable time to the study of the map, and has located his position. He now compares the map with the terrain. He knows that this may save him many mistakes, and he knows that it is never a wasteful procedure.

After comparing the terrain with the map, locating landmarks and choosing a tentative route, Cpl. Thomas then divides the ground into sectors with a recognizable landmark in each sector. He rules out all paths, trails, and roads as covered by the enemy, and picks out the most difficult and wildest route. In these areas, enemy cover is unlikely. He realizes that the sun is in front of him but to make the best of the situation he darkens his face, hands and clothing in order to blend more fittingly into the background. He knows that to escape detection he must use every ditch or depression, and every dark and shadowed piece of terrain.

TECHNIQUE:

At 1830, Cpl. Thomas leaves his observation post near Hillpert Farm and proceeds north about 800 yards through the woods without any trouble. As he takes cover and observes around a rock, he sees a herd of sheep to his front. He does not want to disturb them because they would indicate to anyone who may be watching that something is wrong. Cpl. Thomas sees that it would be easy to go below them and keep out of sight.

Before he proceeds, however, he checks on the direction of the wind by wetting his cheeks. The cheek is extremely sensitive to moisture and wind. The wind, he finds, is coming from the Northeast, so it should be safe to pass below the sheep who have a keen sense of smell. After he has gone below the sheep and is well past them, Cpl. Thomas checks the terrain with the map and finds that his detour was too long and that if he were to continue he would walk into the center of a road net, a point where German guards are certain to be.

The Corporal, heading East, crosses a small stream and begins to climb. Nothing more happens until he is within 200 yds of a big rock. Suddenly, from some place behind the ridge, two crows rise and fly quickly about him calling as they go. Immediately, the Corporal is suspicious.

He knows that a good scout is always suspicious of disturbed animals or birds until he learns the cause of their alarm. The Corporal knows that something, probably a man, has frightened the crows. He immediately takes cover and looks around for a line of retreat if it becomes necessary.

Ten minutes later, three sheep come trotting on the skyline; they stop every now and then to look behind them. Since the sheep are still frightened and continue to trot on, Corporal Thomas is even more certain that a man is somewhere on the ridge. If the man were a sheep herder, he would undoubtedly have a dog with him. The Corporal tests the wind, finds it from the right direction and remains concealed.

After ten minutes of cautious waiting, the Corporal continues his journey up-hill toward the rock. The rock proves to be an excellent landmark. From the present side it looks quite different, but its distinctive size enables him to recognize it and locate himself.

From his map he finds that he is now near the Hillpert Farm. Looking around, the Corporal is able to see a good deal of the country he has not seen before. To observe from the top of the rock would give him an excellent view but it would also silhouette him against the skyline. A place in the grass beside the rock, although it restricts his vision, gives him more security.

He looks around very slowly; he knows that quick movements are easily seen. To his rear Corporal Thomas sees a bright flash several hundred yards above him. At first he is quite dazzled, but soon he recognizes the glare—a man with field glasses. The Corporal thinks to himself, "It is lucky that the observer does not know that he has to shade his glasses to keep them from flashing a message over the hillside."

The observer seems to be looking directly at Cpl. Thomas, but Thomas does not move. His safety lies in perfect stillness. After a few minutes, the observer gets up and starts walking toward the Sabillasville-Thurmont road.

The Corporal continues his observation and makes a mental note to avoid all houses on the way. One of them might have a dog that will bark and give him away. He remembers that the same warning applies to farmyards: there is always some animal that will make a noise. Alarm noises of domestic animals are better known than those of wild animals. Too many people recognize them for what they are.

At this point Thomas sees a shepherd only 100 yards away. Since it will soon be dark, Thomas does not want to waste any time. Keeping his head down, he crawls on his stomach about thirty yards to a deep ditch, down which he is able to make his way unobserved.

After 300 or 400 yards the Corporal stops for a rest. His crawling and rapid withdrawal have tired him. He crawls into some brush, keeps his carbine close by and spends the last half-hour of light observing the countryside.

To the left and below him, the Corporal sees a small stream, which he knows leads to a road and joins the main stream. His plan is to get to the stream and move down the road in the water. The high bank will hide his body and the noise of the water will mask the sound of his movements. If this were still water, he would avoid it as a plague because still water, like dead wood and leaves, is a sound trap. The stream, he knows, flows rapidly because of the sharp decline, so he proceeds to the stream and follows it to the road.

The stream leads to a culvert too small for him to crawl through. He must, therefore, cross the road. His point of crossing gives him a clear view of the road. He sees that the road is gravel and he cannot move either way because there is a guarded road junction several hundred yards on each side. He must then wait for a counter-noise.

After a wait of a half hour a car finally comes by, and the Corporal, the sound of his movements covered by the car noise, dashes across the road to the grove on the other side. As soon as he is again ready to move forward, the door of a house to the right of the grove opens and an enemy solider steps out. Cpl. Thomas, knowing that the soldier will be unable to see clearly for some time because he has stepped from a lighted room into the darkness, quietly leaves the immediate vicinity, continuing on through the trees.

The Corporal then gets back into the stream—he is still too far from the lines to risk leaving footprints in the soft mud banks of the creek.

He follows the stream along the woods, and cuts through the woods to the junction of the creek leading to Sabillasville. The first 500 yards along this creek are easy enough, but the creek bed becomes so shallow and rocky that Thomas must move forward on the ground.

He moves forward through the high grass, stepping high and making each step very carefully. This avoids the brushing noise of the foot moving through tall grass and often keeps the man from tripping over low obstacles.

In this way Cpl. Thomas reaches Sabillasville at 2015. He waits in the shadows of the designated building on the outskirts until he is joined by two other surviving members of his patrol. Following previously laid plans the three men return to their lines.

NOTE

This lecture can be elaborated on and localized to make it more instructional. The personalized type of presentation of fieldcraft is particularly good as it will hold interest. Individual points can be discussed during the lecture as they come up in the hypothetical journey. The story serves to tie them together.

SUGGESTED LECTURE ON
RECONNAISSANCE PROCEDURE
(For the Scout or "Sneak" Patrol)

1. *INTRODUCTION:*

You must gain information about the enemy without arousing his suspicions, and you must get it back in time to be of use. This means that you must move as rapidly as is necessary to get your information back in time, and you must be undiscovered while on reconnaissance. You must leave no trace of your visit. Everything in the zone concerned is your enemy—animals, birds, vegetation, snow, mud—as well as the Japanese or German soldier.

Your job is to avoid changing in any way the terrain being reconnoitered. You must see without being seen, hear without being heard. The best way to do this is to plan your reconnaissance as if the terrain involved were a stage, and your enemy the audience. Learn to see everything as he would see it.

Whenever possible, plan your route so that the terrain will aid you in concealing yourself, maintaining direction, and finding your objective. Use you map, but if at all possible, actually observe the country you are to move through. Observe it from a safe, concealed spot, from which you can see without being seen. If you use field glasses, be careful that the direct rays of the sun are not allowed to fall on the lenses, for this will cause a reflection that may give your position away.

Although only direct observation of your terrain will give you a good idea of what you are going to move through, it is best to check it with a map, to fix the ground plan firmly in mind and to find any accidents of the terrain which may be in defilade from your sight. Where maps are not available, make a sketch of all important terrain features and landmarks which can serve to aid you along your route, or use an aerial photo if you can get one.

Landmarks must be objects that can be easily found and recognized. Hills that seem definite during the day will often resolve themselves into vague shadows at night. You can avoid this by picking unmistakable points such as streams, edges of woods, walls, groups of houses, and any particular objects that you will not mistake for others of the same kind. Select objects that can be easily recognized from either side.

Pick the landmarks by bounds, but do not make the bounds too large. At night, when visibility is difficult, bounds should be smaller. If it is possible, determine the azimuths to the points you have selected. You may also guide by the sun and the stars—anything that will keep you on the correct course. Learn the direction of flow of the streams in the area. Take a native guide or a soldier who has previously patrolled the area, if either is available or reliable.

Overlook nothing that will keep you on the right road, save time, or help in any way in the accomplishment of your mission. Remember that any time estimated for daylight reconnaissance will be at least double when working at night. Don't measure distance from your map as the crow flies, and then wonder why you are hours late in returning.

You must spend at least part of the day preceding your patrolling mission in observation of the terrain you are to cover, so that you can pick your route. By this observation you cannot only maintain your direction, but also determine what the enemy is doing, locate his probable strong points and listening posts, and other areas of forced passage that must be avoided. Find out as much as you can about the enemy from the officer in charge of the point from which you depart your lines. You must also determine if the enemy is patrolling the same area, for if he is, you must get there first to avoid any ambush the enemy may lay.

If the terrain is open, plan not to waste too much time near friendly troops, but move swiftly, using long bounds, though on the alert. Each halt should be at a point where you can conceal yourself and retire if necessary. Direction should be checked at these halts. However, do not stop to read your compass every few minutes, for this only serves to hold up your progress. If you can safely do so, use all the daylight hours in the march toward your objective.

Planning is at least half the job of reconnaissance. Your equipment must follow the same principle, that of remaining unobserved. There must be no shiny articles, nothing that produces or reflects light or makes noise, or in any other way attracts the attention of the enemy. All men must be checked physically, because any one who might sneeze or cough will jeopardize the success of the patrol. White handkerchiefs and light leggings, bright clean faces and hands, shiny equipment or white maps will betray you to the enemy. Leave the handkerchiefs and leggings behind; dull the equipment; smudge the hands and face, and cover the map.

Leggings may also betray you by scratching against the underbrush and other hard objects, producing a noise which can be heard for some distance. Half-filled canteens, rattling canteen cups, tinkling identification tags, jingling coins and keys are further sound enemies. The steel helmet will scratch against trees, causing noise, impair your hearing, and be an unmistakable outline. It, too, must be left behind. You want to travel *lightly, tightly,* and *quietly*, so make all adjustments in your equipment and clothing before starting out. Wear comfortable, warm clothing that will blend with the background, both day and night. After all identification has been left behind and the patrol leader has made an inspection, the patrol is ready to leave.

Before starting out, certain specific arrangements for maintenance of direction should be made; for example: points may be designated where artillery shells will be dropped at appointed times. In this way, the patrol can quickly check its orientation by observing where the shells burst. (It is necessary for the artillery to register on them during daylight hours, so that when the shells are dropped for orientation at night, they will hit exactly where they are wanted.)

Each patrol member must have clearly in mind the fact that he must never stray from the formation and cause the leader to lose control. A good procedure is to have the patrol halt every 100 paces for a check-up.

2. *POINT OF DEPARTURE*

When all men are familiar with the entire situation—friendly and enemy— when they know the mission and everything about it that can be of any help to them; when they have been checked for physical fitness and proper equipment and proper preparation of their equipment and themselves, then they move to the point on the outpost line through which they will leave.

At this point they pick up any additional information which is available, check the password, countersign and reply, and make sure that the outpost line will be expecting them when they return. In leaving the last elements of friendly troops, try to pick an unobserved spot, and make the best use of any haze, fog or mist that may be available. Whenever there is a natural phenomenon such as this, it is wise to take advantage of it even if it is earlier than you had planned, because it will give more time for your reconnaissance.

Avoid skylines and hill crests in your movements. Never walk over or stand on a hill so that you will be silhouetted to an enemy observer. When crossing a hill, avoid the crest, try to use the sides, and cross by stages, patrol member by patrol member, on the alert and protected against surprise. *When in doubt, crawl.*

This same procedure holds true for crossing an exposed area, road, clear hillside, field, etc. Try to pick a background that will blend as much as possible with your uniform. When there is a bright moon, cling to the shadows as much as possible.

If you find it necessary to cross a river or a lake, try to find a point where you will be camouflaged by debris. Be careful of the splashing sound of your swimming, especially at night.

If you move along a beach, you will be less likely to be observed from craft offshore if you are near the water's edge where the high spray and foam conceal you. However, if you are more afraid of observation from enemy on the shore, move close to a bank or the rock underside of a cliff.

When moving over soft ground, you will make very little sound if you lower your foot to the ground heel first, then gently place the ball of your foot down. When on hard ground, the procedure is reversed—toe first, then heel. Moving through grass, brush, or any type of high growth you will be more quiet if you raise your foot high enough to clear the growth, then set it down heel first.

When you come to an obstacle, don't bunch up. Follow the assigned procedure of halting when the patrol comes to any terrain feature which it cannot pass without a change of formation. No more than one man at a time should cross the obstacle, and he should be covered by the men who have yet to cross, as well as by those who have already crossed.

The best procedure for movement at night is to remain low. If you stay among the lowest features of the terrain, you will be at an advantage, because looking upward, you can pick up any movement which takes place, against the background of the sky. An enemy soldier on the hills cannot easily see you, because he is looking down into the darkness of the stream bed or draw. Remember that the sky always appears just a little lighter than the ground at night.

In relation to stream beds, you will find that they make good avenues of movement for patrols. The noise of the moving water covers you, and you leave very little trail. However, you must be cautious because you may work yourself into an indefensible low point which is covered by fire and enemy mines. Good reconnaissance and observation during the daylight hours prior to moving out on the patrol will, of course, help avoid such situations.

Any route that is an "obvious" one must be avoided. Do not follow any route that moves like a highway into the enemy area, because it will most surely be covered by fire or mines, or ambushed. Neither should you follow a route which will force you to move into a position from which you cannot escape if the enemy fires on you.

Avoid trails in the vicinity of the enemy. Also avoid roads, houses, and farm yards. Be especially careful of farm yards, because domestic animals will expose your movements. Do not go near isolated points on the terrain, for these are very easily kept under observation by the enemy, and any movement near them is usually obvious.

Do not be disturbed by rain. In fact, you should welcome the sight of it, because it covers your movement very well. Rain, hail, or any type of stormy weather adds to the success of a patrol, because the enemy is less alert, your noise is covered, and your trail obliterated.

Try to avoid making trails or tracks. This is especially necessary if you expect to remain in the area for any time, or if you must return by more or less the same trail. Your trail, in any case, tells the enemy that you were there, and that is important information to him.

Here a conflict often arises, the question being: "Shall I make a trail or a noise?" Tall grass conceals you and may be quiet, but it leaves a trail. Many times you will have such a decision to make, and it is one that can only be made after consideration of the old standbys, the situation and the terrain. Sometimes you can make use of the trails of animals.

3. *APPROACH TO OP OR OBJECTIVE*

As you approach your objective, make use of every bit of cover and concealment you can find. Walls, rocks, brush, even grass can be your friend at this point. Never break the smooth outline of any natural feature such as a large rock or a skyline by putting any part of your body over it. Move by bounds, and lie still between each bound so that you may listen. Never make moves so quickly that you draw observation to yourself.

The wind is important when you reach a point near the enemy. Determine its direction by whatever means you can. It has been suggested that wetting the cheek will do the job, since the cheek is very sensitive. Be sure to remain downwind, so you will not be troubled by sounds and scent moving down to the enemy. On the other hand, enemy noise will come to you, and you can pick up more information.

If you make a noise, stop. A sound that is not repeated during a few minutes will not alarm most sentries. Move under cover of other normal noises, MG fire, airplanes, or cars. When you are near the enemy, most of your time will be taken up waiting to hear if he has heard you.

When you are near the enemy, you will be watching and listening so intently that you may see and hear things which do not actually exist. Be careful of this. Stars may turn into lights, the shadowy outline of a bush may become a sentry. In instances like these, you must exercise all your self-control. You are never captured until they have actually discovered you—so if a light shines on you, lie still. If you don't move, chances are that you will remain unseen. If a burst of fire strikes near you, don't take off in a panic; nine chances out of ten, the fire is merely directed at a sound, and not necessarily at you.

Crawl at night, feeling for what is in front of you, but use a light touch, for there may be booby traps or pressure mines concealed.

If you are approaching an observation point to be used during the coming day, have a scout proceed under protection of the patrol to reconnoiter for presence of the enemy. Use a scout for any area of which you are not certain.

An observation post should be picked not only for visibility and concealment, but also checked for a possible line of retreat and self-defense. Camouflage yourself during the morning haze, and make all adjustments which are necessary for the day's observation. Have everything necessary within easy reach.

The ability to lie still for long periods is definitely necessary for observation. This is the best protection against any enemy who might see you.

Do not select large, obvious places for your OP. They are sure to be suspected by the enemy. Use low brush, jagged rocks, or any other objects which are unimportant in themselves, that will give you concealment either within them or in their shadows. You can also add to the effect by camouflaging yourself: paint your face, camouflage your helmet, or use a camouflage suit. Never "blackface" as that is as bad as no camouflage.

Pick the small, inconspicuous objectives and learn to lie still. In a recent operation, a reconnaissance sergeant lay all day without moving in a clump of brush three feet high, at 300 yards from the enemy division headquarters, on the side of the hill facing that headquarters. Having aroused no suspicion, from dogs, civilians, guards, or others, all normal life passed by and around him. This, of course, is over-risky, and not to be recommended, but it does show what can be done by learning to lie still.

You should lie prone in all cases of doubt. Keep as close to the ground as possible. To observe, move your head slowly and steadily, avoiding any abrupt movements that would disclose your position. Observe from one side of the area to the other, in belts of terrain which overlap as they recede.

If you can find a well-concealed place to sit, do so without moving your hands, as this is one of the quickest giveaways of even the best camouflaged groups. Keep your hands still.

The sergeant mentioned above also tells the story of sitting and observing the enemy rear from a position high on a hill in a region about 250 yards from a farm house. He was situated between two jagged rocks under a small bare tree. A patrol of enemy came up to the farm house, searched it, and then turned and advanced directly toward him. He sat perfectly still, so still that they came up to a point 100 yards from him, and then followed a path which led away from him. If he had moved, he would have given himself away.

4. *OBSERVATION*

An observer's duty is one of the most important duties in combat. By using his eyes and ears, he is the one who completes the picture of the enemy, giving strength, organization, and possibly even plans; all of those things which are essential for a commander to know before he can make a decision.

Even a trained observer can observe only part of the total area before him. This will, therefore, be most effective if his attention is directed to certain specific types of information in specific areas. Observers are sent out from all different units, not only at the front, but also at the flanks and rear. They are sent out in order to guard against attack by air, tanks, and parachute troops, and in some countries against disloyal inhabitants. If an observer sees something which may not be of use to his own unit, he must realize that it may be of great value to other units. Therefore, all information of military value should be reported whether or not it appears to be of direct use for the observer's own unit.

It is better to report too often than too seldom. Negative information is of value. The fact that all is quiet in a certain area or section may be of great value to the commander. Areas denied to ground observers by distance, difficult terrain, or hostile ground action should be observed by employing aerial observation.

When you have a certain sector to observe, divide the terrain into a series of zones. The nearest zone takes in the ground just beyond the front line of your unit, while the zone farthest away includes the limiting time in depth of your section. The observer begins with the zone nearest him. He then goes on searching the terrain from one side to the other, progressively, by zones.

You search not only for enemy movements, but for indications of enemy activity, such as trenches, paths, gun positions, wire entanglements and observation posts.

Once you have determined enemy activities, locate them on a map of the area. If no map is available, draw a sketch and locate the information on it. Then send this information back to your commanding officer.

The ability to estimate the organization, strength and condition of the enemy's troops and get this information back to your commander may result in the proper and early disposition of your unit to defeat the enemy. You should be as familiar with the organization of the enemy's forces as you are with your own. By observing troops of our own army in camps and on maneuvers, you will become familiar with their strength and composition. You should know their road space, the front they cover in deployment, and their appearance under varied conditions. By so doing, you will learn to estimate the strength of similar enemy units under similar conditions.

You may estimate the strength of a column on the march by noting the time required to pass a given point. For example: the infantry, in a column of threes, occupies .8 of a yard per man; therefore, there are 125 infantrymen in such a column 100 yards long. The cavalry, in a column of fours, occupies 1 yard per man; therefore, there are 100 cavalrymen in a column of fours 100 yards long. On the average, an infantry unit of 110 men will pass a point in one minute in a column of threes. One hundred ten cavalrymen at a walk in a column of fours and five horse-drawn guns or caissons at a walk will also take one minute to pass a given point. The number of vehicles in a motorized or tank unit to pass a given point varies with the speed of the column and the distance between vehicles.

Lights, fires, smoke, dust or noise may give information as to the strength, composition and actions of an enemy force. Information of the enemy frequently may be gained by a study of tracks. A familiarity with enemy organization and equipment aids you in making an accurate estimate of the composition of enemy forces by noting various tracks. The following points will aid you in acquiring information from tracks:

1. Different armies, and sometimes different organizations in the same armies, wear different kinds of footwear.

2. Large columns wear a dry road smooth and flat. Comparatively little dust will be deposited on the roadside vegetation after the passage of foot troops, whereas a great deal of dust is stirred up by mounted troops, trucks, and tanks.

3. Small-wheeled tracks indicate the passage of MG's, antitank guns, mortars, motorcycles, or other small reconnaissance vehicles. (Tire tracks may also be distinctive.)

4. Artillery, tank and supply columns make very distinct tracks. The passage of diesel powered tanks and other vehicles can sometimes be detected by the distinctive, persistent odor of diesel fuel.

5. An indication of condition and morale of troops can be found by the following points: the distance between hourly halts may indicate the rate of march; ground cleared up after the halt indicates good discipline; rubbish, packs, rifles, and ammunition scattered about are an indication of poor discipline; tracks leaving a column in the direction of an orchard, farmhouse or well also show poor discipline; a hasty withdrawal or route is reflected in heaps of enemy stores and materials in good condition; burned supplies and destroyed materials indicate a more orderly withdrawal.

6. Speed and direction of the vehicle are shown by the following: a car passing through mud or water will show wet tracks on the side on which it leaves the mud or water. Mud and water are scattered more by a swiftly moving car than by one moving slowly; a swiftly moving car scatters piles of dirt and sand, whereas a slowly moving car leaves a deep, smooth track. A wheel going over a hole in the ground leaves a deeper mark on the side toward the direction of travel—the greater the speed, the deeper this mark will be.

5. *RETIREMENT AND RETURN MARCH*

In retiring from an OP, or in the return from reconnoitering an objective, the same precautions must be observed as were used in advancing. If you crawled up, crawl back. Until you are sure it is safe, do not crouch or walk.

Do not hurry! Too many patrols or scouts, in the return trip, act as if their mission were over; remember, you must return with the information without letting the enemy know you have it. Only in the case of a fighting patrol can you violate this rule.

Keep all the cover you can between yourself and enemy observation. Be silent and vigilant against possible ambush. During the relief that is felt by the men on a return march, there may be a tendency toward whispering, smoking, scuffing of the ground, and carelessness. A fighting patrol or raiding party may get away with it; a reconnaissance patrol will ruin its mission. It is definitely wrong in either instance, and will almost invariably lead to casualties.

You may be able to use more defensive formations, and proceed by longer bounds than you would ordinarily make in the advance. However, all ambush points such as fords and forced points of passage must be checked carefully.

If a man has been lost, be on the alert and don't fire until you recognize every one you meet. If, in returning, you are a lost or very much delayed patrol, watch out for other friendly patrols. Many men have been killed in battles between friendly patrols. Be sure you know the password and identifications of friendly patrols.

On returning to your own lines always take the route previously decided upon. To return by the same route used going out courts enemy ambush. Upon reaching the lines, have one man approach with the countersign, always under the protection of the patrol. Do not bring the entire patrol in until contact has been established. There is always the chance that a nervous sentry will fire, although the outpost is expecting you.

On a fluid front, never take it for granted that the line, village, or hill that you left two days ago is now friendly. Make sure that it is still friendly or you may walk into an ambush. The front may have changed, positions may have been abandoned, or the troops may have retired temporarily. Always have a reserve of strength in order to reach a goal beyond your estimated safe line. Your entire trip, if well planned and organized, should fall within your estimated time bounds, but night, loss of contact, or loss of direction all conspire to throw the best schedules out of joint.

If you separate or become lost, do not wander in panic. Move to temporary safety and stop. Remain cool and try to establish your direction. Then go to the rendezvous point. If you have been pursued, in your panic you may run into more of the enemy, which has happened often. If it is near dawn, pick cover for the following day and remain there all day. If you outwit the other fellow, you will be the winner.

Three men on reconnaissance, well within the enemy lines, were ambushed when returning from patrol. Daylight was almost upon them. The only cover was a small freshet in a deeply cut valley under some small bushes. Cavalry patrols, guards, and casual soldiers passed them all day. It was such an impossible place to hide that no one looked for them. Some of the enemy even washed in the stream above them. It was cold and uncomfortable sitting in the cold water. They did not move although they often thought they were seen. Had they moved, they would have been discovered.

In battle, many patrols well camouflaged and in good position have been betrayed by moving an arm or shifting the body. When you think you are under enemy observation and flares appear, freeze in position. You can realize the efficacy of such action by trying to find a man at night who is standing motionless 100 yards from you.

6. *SPECIAL OPERATIONS*

Your mission will often require you to pass through and work behind the enemy outguards. To do this you must be able to pass through enemy wire and cross trenches quietly.

To cut a gap in wire requires time and may alarm the enemy. When possible, walk over the low bands and crawl under the high bands. To step over low wire, crouch low, so that you can see the strands against the sky. Grasp the first strand with one hand, and with the other hand, reach forward and feel for a clear spot for your feet. To go under wire on your back, grasp the lowest strands and hold them clear of your body while you slide under. Never to be forgotten is the fact that barbed wire entanglements frequently contain antipersonnel mines. Before crossing an entanglement in front of the enemy position, you should examine it as carefully as possible against the skyline. In this way you may be able to locate any small wires which may be connected to antipersonnel mines or to tin cans filled with pebbles used to spread the alarm. Most antitank mines are not detonated by the weight of a man, and may be crossed on foot without special precaution—unless, of course, the field is protected by antipersonnel mines. Antipersonnel mines that are operated by tripwires are best passed at a point where the ground is smooth and hard and there are few or no bushes. Lift your feet and set them down as if you were passing through high grass.

Many times it will be necessary to cut wire in order to get through it. If you are working alone, cut it near a post. Grasp the wire in close to the post and cut it between your hand and the post. In this way you will be able to muffle the sound and keep the loose wire from noisily snapping. Bend back the loose end to form a passage. If another scout is working with you, one of you holds the wire close to the wire cutters, in order to muffle the sound and prevent the loose ends from flying back, while the other cuts. In cases where it is necessary to cut electrically charged wire, specially insulated wire cutters and specially trained personnel are required. Remember that pressure devices on booby traps can be placed on tight wire so that they explode when it is cut.

Reconnaissance procedure follows certain general rules. Under certain conditions, however, these rules are modified by special terrain and climatic features, such as jungle, desert, or snow. Clothing, weapons, and methods are adapted to achieve the same ends as in ordinary reconnaissance. In desert warfare, dust storms, wind, mirages, and moonlight all affect visibility.

In snow and extreme cold many parallels to desert reconnaissance can be drawn.

LECTURE ON SCOUTING AND PATROLLING USING REGULATIONS OF ROGERS RANGERS AS A BASIS

The old adage that history repeats itself is applicable to principles of warfare. Basic military tenets repeat themselves from one war to another, from one century to the next. Pertinent are the principles of scouting and patrolling which have remained unchanged for over centuries—despite revolutionary changes in weapons and techniques of modern warfare.

We can draw amazingly similar parallels between our present day scouting doctrines and those of military forces fighting centuries ago. An outstanding example of this is the regulations for Rogers' Rangers of the early 18th Century, which closely correspond to the rules laid down for our scouts, our patrols—our present day Rangers. In no way do they differ; they only serve to reaffirm basic principles of scouting that will always prevail.

Rogers' Rangers was a force of militia conceived and nurtured in this country by Robert Rogers during the Seven Years War from 1756 to 1763. Made up of frontiersmen trained to fight the French and Indians, they incurred a great reputation for their successful military accomplishments and indomitable courage and endurance. Adopting Indian tactics, the Rangers were a sort of magnified combat-*reconnaissance* patrol. They distinguished themselves in campaigns around Lake George, Quebec, Montreal and Detroit. Aside from the fine caliber of leadership and men, the secret of the Rangers was in their training and methods of fighting. Based on sound principles, their methods are the methods our scouts and patrols should use today.

Presented in the following section are the regulations for Rodgers' Rangers, along with pertinent comments as observed by the author.

I. *"All Rangers are to be subject to the rules and articles of War: to appear at roll call every evening on their own parade, equipped, each with a firelock, sixty rounds of powder and ball, and a hatchet, at which time an officer from each company is to inspect the same, to see they are in order so as to be ready on an emergency to march at a minute's warning; and before they are dismissed, the necessary guards are to be draughted and scouts for the next day appointed."*

Here is typified the emphasis placed on patrol inspection today. Equipment is checked with regard to noisiness, appearance, and suitability. Nothing that will give the patrol away is permitted—loose mess kits, rattling canteens, dog tags, or loose change. Camouflage principles are applied in keeping the equipment and clothing concealed from enemy eyes. Darkening with paint, dirt or grease eliminated the danger of light reflection. Arms—rifles, tommy guns, or automatic rifles—are tested. Preparation, as with the Rangers, is still a keynote of all patrolling. Preparation in knowing the mission, the terrain and the enemy before starting out on the patrol. Preparation in being ready for any emergency that may arise.

II. *"Whenever you are ordered to the enemies' forts or frontiers for discoveries, if your number be small, march in a single file, keeping at such a distance from each other as to prevent one shot from killing two men, sending one man or more forward, and the like on each side, at a distance of twenty yards from the main body, if the ground you march over will admit of it, to give the signal to the officer of the approach of the enemy, and of their number, etc."*

The maintenance of proper intervals cannot be overemphasized. In bivouac, on the march, or on patrol it plays an equally important role. In training it must be drummed into the men time and time again, or else in combat it will be learned too late. It is human nature for men to bunch up at obstacles, but they must be taught and they must learn to keep at such a distance from each other as to prevent one shot from killing two men. Point and flank security must be maintained at all times, and the use of a column formation is basic especially in dense woods. The formation must enable complete control.

III. *"If you march over marshes or soft ground, change your position, and march abreast of each other to prevent the enemy from tracking you (as they would do, if you marched in a single file) till you get over such ground, and then resume your former order, and march till it is quite dark before you encamp, which do, if possible, on a piece of ground that may afford your sentries the advantage of seeing or hearing the enemy some considerable distance, keeping one-half of your whole party awake alternately through the night."*

Nothing gives away a scout's presence more unnecessarily than his tracks. He may camouflage himself, and he may move silently, but if he leaves footprints or beaten down brush he no longer conceals his presence from the enemy. Moving over sand, the scout should carefully smooth his tracks, either with a branch or with leaves. Moving over fields, he is careful not to disturb twigs or bushes. Roger's ruse of many tracks to cover up one is a good one only if a large force and not a patrol is advancing. Often soft ground cannot be avoided, and by marching abreast a force will make it difficult for the enemy to pick up one particular track. "All around" security at a halt must always be maintained by the patrol.

IV. *"Some time before you come to the place you would reconnoiter, make a stand, and send one or two men in whom you can confide, to look out the best ground for making your observation."*

The scout is always sent out to reconnoiter dangerous points before the patrol is committed. The principle of "scouts out" is a basic one, and too often neglected. Failure may lead to the whole patrol's being ambushed. Before the patrol is committed to a dangerous area, the patrol leader himself usually goes out to reconnoiter, but at all times the use of the scout must be constant.

V. *"If you have the good fortune to take any prisoners, keep them separate, till they are examined, and in your return take a different route from that in which you went out, that you may the better discover any party in your rear and have an opportunity, if their strength be superior to yours, to alter your course, or disperse, as circumstances may require."*

Two important principles are mentioned here. Prisoners must immediately be segregated. Interrogation is facilitated by keeping them apart, otherwise the officers and non-coms will tip off the privates as to what and what not to say. Combat-reconnaissance patrols may take prisoners, and little or no information will be forthcoming if prisoners are allowed to mix and talk to one another.

The second principle is that of patrols returning by a different route. Patrol experience has shown that if the enemy once discovers patrols are in the habit of using the same approach, he will lay an ambush. Returning by a different route serves the added purpose of permitting the patrol to seek out other enemy dispositions and to acquire better knowledge of terrain.

VI. *"If you march in a large body of three or four hundred, with a design to attack the enemy, divide your party into three columns, each headed by a proper officer, and let those columns march in single files, the columns to the right and left keeping at twenty yards distance or more from that of the center, if the ground will admit, and let proper guards be kept in the front and rear, and suitable flanking parties at a due distance as before directed, with orders to halt on all eminences, to take a view of the surrounding ground, to prevent your being ambushed, and to notify the approach or retreat of the enemy, that proper dispositions may be made for attacking, defending, etc. And if the enemy approach on your front on level ground, form a front of your three columns or main body with the advanced guard, keeping out your flanking parties, as if you were marching under command of trusty officers, to prevent the enemy from pressing hard on either of your wings, or surrounding you which is the usual method of the savages, if their number will admit of it, and be careful likewise to support and strengthen your rear guard."*

Security at all times: another cardinal tenet. Flank guards and rear guards to guard against surprise both on the march and at the halt. Surprise is a very effective weapon, but security is just as effective a countermeasure. Too often, patrols, careful to provide security during the advance, become lax and fail to provide all-around guards when they halt or when they are returning. Rear security is as important as security to the front and to the flanks, since patrols operate in enemy territory and the enemy is likely to appear from any direction.

Movement by bounds continues to be a basic of patrol action. Frequent halts to determine the next bound and to listen for enemy sounds is an accepted patrol procedure. Keeping the flank guards well out lessens the chances of ambush or of being surrounded "which is the usual method of the savages"—both ancient and modern.

VII. *"If you are obliged to receive the enemy's fire, fall, or squat down, till it is over, then rise and discharge at them. If their main body is equal to yours, be careful to support and strengthen your flanking parties to make them equal to theirs, that if possible you may repulse them to their main body, in which case push upon them with the greatest resolution with equal force in each flank and in the center, observing to keep at due distance from each other, and advance from tree to tree, with one-half of the party before the other ten or twelve yards. If the enemy push upon you, let your front fire and fall down, and then let your rear advance through them and do the like, by which time those who before were in front will be ready to discharge again, and repeat the same alternately as the occasion shall require; by this means you will keep up such a constant fire, that the enemy will not be able easily to break your order or gain your ground."*

This early conception of fire and movement still serves as the basis for a tactical advance. Even the smallest units—squads—employ the principle. One squad takes cover and engages the enemy with fire while the other squad manuevers to the enemy's flank to get enfilade fire or to take the enemy by surprise.

VIII. *"If you oblige the enemy to retreat, be careful in your pursuit of them, to keep out your flanking parties, and prevent them from gaining eminences, or rising grounds, in which case they would perhaps be able to rally and repulse you in their turn."*

To be ever alert for enemy counter-attacks is as much a requisite now as it ever was. Counter-attack is the basis of German defensive tactics. The capture of a German stronghold inevitably means a counter-attack, and it must be strongly guarded against. Gaining ground from the enemy means only temporary victory unless our attacking forces can throw up a defense as strong as the attack. Combat-reconnaissance patrols must understand such tactics.

Alleviation of the danger can be enhanced by always maintaining contact with the enemy. Just as the Rangers kept contact with the Indians, so today we must always maintain contact by patrols with the forces of the Germans or the Japanese. They must be kept in visual view by our forward elements—both in the attack and in the defense—otherwise, they can surprise us and we are less able to anticipate, and subsequently take precautions against them.

IX. *"If you are obliged to retreat, let the front of your whole party fire and fall back, till the rear hath done the same, making for the best ground you can; by this means you will oblige the enemy to pursue you, if they do it at all, in the face constant fire."*

If a retrograde movement is unavoidable, a delaying action should be undertaken. Delay the enemy by smoke, fire, or other means, while your main body makes an orderly retreat. "Oblige the enemy to pursue you *in the face of a constant fire."*

X. *"If the enemy is so superior that you are in danger of being surrounded by them, let the whole body disperse, and every one take a different road to the place of rendezvous appointed for that evening, which must every morning be altered and the whole party, or as many of them as possible get together, after any separation that may happen in the day; but if you should happen to be actually surrounded, form yourselves into a square, or if in the woods, a circle is best, and if possible make a stand till the darkness of the night favors your escape."*

Sneak patrol members should be instructed to disperse in the event of ambush. Each man moves away individually, meeting at a prearranged rendezvous point. Progressive rendezvous points are designated along the route of march to cope with any situation that may arise.

XI. *"If your rear is attacked, the main body and flankers must face about to the right and left, as occasion shall require, and form themselves to oppose the enemy as before directed; and the same method must be observed, if attacked in either of your flanks, which means you will always make a rear of one of your flankards."*

This regulation can be interpreted in present day context as: be ever on the alert and watch out for surprise attacks, particularly from the rear. Patrols must be mobile and flexible enough to be able to protect themselves from an attack from any direction.

XII. *"If you determine to rally after a retreat, in order to make a fresh stand against the enemy, by all means endeavor to do it on the most rising ground you come at, which will give you greatly the advantage in the point of situation, and enable you to repulse superior numbers."*

The success of a patrol depends upon the proper use of terrain. All patrol members must know the terrain and how to use it. The effectiveness of Rogers' Rangers was based on such an understanding. They knew that a force which could locate itself on commanding ground had a superiority, in regards to both the firefight and observation, and they knew how to fit themselves into the terrain in order to achieve such a position. They understood the lessons that we have learned at Cassino and elsewhere—namely, that a force, inferior numerically, but well entrenched on a commanding piece of terrain, can repulse a larger attacking force.

XIII. *"In general, when pushed upon by the enemy, reserve your fire till they approach very near, which will then put them into the greatest surprise and consternation, and give you an opportunity of rushing upon them with your hatchets and cutlasses to the better advantage."*

The practice of withholding fire until the enemy approaches very close is an old one. The command "Don't shoot until you see the whites of their eyes" has echoed through all wars since. Today our rifle platoons are taught to withhold fire until the enemy is 400 yards away. An enemy patrol is more effectively eliminated and the element of surprise is stronger in an ambush if he is engaged at close quarters.

XIV. *"When you encamp at night, fix your sentries in such a manner not to be relieved from the main body till morning, profound secrecy and silence being often of the greatest importance in these cases. Each sentry, therefore, should consist of six men, two of those whom must be constantly alert, and when relieved by their fellows, it should be done without noise; and in case those on duty see or hear anything, which alarms them, they are not to speak, but one of them is silently to retreat, and acquaint the commanding officer thereof, that proper dispositions may be made; and all occasional sentries should be fixed in like manner."*

Security at night is important, but secrecy and silence often are even more so. Silence is the reconnaissance patrol watchword. A patrol is silent on the move; silent, at the halt. Sneak patrols are formed on the principle of silence. If the enemy cannot hear you, he will not have cause to look for you. Outguards at a halt must be as silent as a patrol on the move. Reliefs must be executed in silence. If sentries sight the enemy, they must continue to remain silent, for although you can see the enemy, he cannot always see you. The use of the cossack type sentry post is still effective.

XV. *"At the first dawn of day, wake your whole detachment, that being the time when the savages choose to fall upon their enemies, you should by all means be in readiness to receive them.*

Alertness at dawn—that is stressed in all phases of combat. H-hours of today's battles often come at dawn. The enemy can attack at dawn as well as we can. Preparation for an attack at dawn, whether one is expected or not, is required by all units in combat. Laxness at any time of day is inexcusable, but laxness at sundown or in the early daylight hours may be fatal. Patrol actions are often best performed at dawn or sunset.

XVI. *"If the enemy should be discovered by your detachments in the morning, and their numbers are superior to yours, and a victory doubtful, you should not attack them till the evening, as then they will not know your numbers, and if you are repulsed, your retreat will be favored by the darkness of the night."*

An attack at night, rather than during the day, is always advisable for a force inferior in numbers to the enemy. Night covers up numerical deficiencies and excessive firepower can make up for it. At night, a combat patrol can do considerable damage to a large unit despite its smaller size. Should the attack be repulsed, again night favors a more successful withdrawal.

XVII. *"When you stop for refreshment choose some spring or rivulet if you can, and depose your party so as not to be surprised, posting proper guards and sentries at a due distance, and let a small party waylay the path you came in, lest the enemy should be pursuing."*

XVIII. *"If in your return, you have to cross rivers, avoid the usual fords as much as possible, lest the enemy should have discovered, and be there expecting you."*

Reconnoitering a stream before crossing it, and having security out while crossing it is important. A pair of scouts should move upstream and down, looking for a suitable fording place and watching for signs of the enemy. Once a good place is found, one scout crosses the stream first, while the rest of the patrol remains concealed, ready to protect him by fire. The scout reconnoiters the other side, and if it is clear, signals back to the patrol. Patrol members then cross the stream one at a time. Flank and rear security is alert throughout.

XIX. *"If you have to pass by lakes, keep at some distance from the edge of the water, lest, in case of an ambuscade, or an attack from the enemy when in that situation, your retreat should be cut off."*

The principle here involves keeping away from traveled routes and likely points of ambush and enemy attack. Experienced patrols will always keep away from lake's edges, as they will from edges of woods. By keeping away from lakes, chances of being cut off are minimized. By avoiding the edges of woods in the daytime, patrols avoid being seen.

XX. *"If the enemy pursue your rear, take a circle till you come to your tracks, and there for an ambush to receive them, and give them the first fire."*

Give the enemy a taste of his own medicine. If a patrol suspects it is being trailed, it can make a feint, and attempt to circle around and ambush the enemy or it can have part of its members lay in wait for pursuing the enemy.

XXI. *"When you return from a scout and come near our forts, avoid the usual roads, and avenues thereto, lest the enemy should have headed you, and lay in ambush to receive you, when almost exhausted by fatigue."*

Here again is the principle of avoiding traveled routes, and returning by a different route from the route of approach.

XXII. *"When you pursue any party that has been near our forts or encampments, follow not directly in their tracks, lest you should be discovered by their rear-guards, who, at such a time, would be most alert; but endeavor, by a different route, to head and meet them in some narrow pass, or lay in ambush to receive them when and where they least expect it."*

Don't follow the enemy blindly, give him credit for being as smooth as you are. Lay an ambush in a place where he is canalized or pick a place where the element of surprise will be used to its fullest extent.

XXIII. *"If you are to embark in canoes, battoes or otherwise, by water, choose the evening for the time of your embarkation, as you will then have the whole night before you to pass undiscovered by any parties of the enemy, on hills or other places, which command a prospect of the lake or river you are upon."*

Travel by night when in enemy country. Doing so prevents his observation from being effective. Night patrolling missions are more frequent than day when contact has been established.

XXIV. *"In paddling or rowing, give orders that the boat or canoe next the stern most, wait for her, and the third for the second, and the fourth for the third, and so on, to prevent separation, and that you may be ready to assist each other in any emergency."*

The procedure to maintain control and contact is the same as that used in night patrolling in dense undergrowth. Each man in the column formation can be made responsible for maintaining contact with the man behind him.

XXV. *"Appoint one man in each boat to look out for fires, on the adjacent shores, from the numbers and size of which you may form some judgment of the number that kindled them, and whether you are able to attack them or not."*

Know your enemy and his organization so that when his dispositions are observed, his strength can be better determined. Maintain constant observation.

XXVI. *"If you find the enemy encamped near the banks of a river or lake, which you imagine they will attempt to cross for their security upon being attacked leave a detachment of your party on the opposite shore to receive them, while, with the remainder, you surprise them, having them between you and the lake or river."*

Utilize the element of surprise and natural terrain features. Figure out the enemy's probable lines of action and act accordingly.

XXVII. *"If you cannot satisfy yourself as to the enemy's number and strength from their fire, etc., conceal your boats at some distance, and ascertain their number by a reconnoitering party when they embark or march, in the morning, marking the course they steer, etc., when you may pursue, ambush and attack them, or let them pass, as prudence shall direct you. In general, however, that you may not be discovered by the enemy on the lakes and rivers at a great distance, it is safest to lay by, with your boats and party concealed all day, without noise or show, and to pursue your intended route by night; and whether you go by land or water give out parole and countersigns, in order to know one another in the dark, and likewise appoint a station for every man to repair to, in case of any accident that may separate you."*

A larger combat patrol is frequently used as a base of operations. Smaller sneak patrols are sent out from it to reconnoiter the enemy.

It is often advisable to lay up during the day and make all movement at night. Modern means of recognition involves the use of challenge, password, and reply so as to provide additional means of recognition.

Rendezvous points that are progressive as the patrol advances must always be selected.

XXVIII. *"Such in general are the rules to be observed in the ranging service; there are, however, a thousand occurrences and circumstances which may happen, that will make it necessary, in some measure to depart from them, and to put other arts and strategems in practice; in which cases every man's reason and judgment must be his guide, according to the particular situation and nature of things; and that he may do this to advantage, he should keep in mind a maxim never to be departed from by a commander, viz: to reserve a firmness and presence of mind on every occasion."*

In patrolling, nothing is stereotyped. Terrain, enemy and local conditions will always affect the situation. On occasion, cardinal principles of patrolling will have to be violated to fulfill the mission. The judgment of the leader and his initiative must be relied upon in these instances. In the last analysis the use of experience, common sense and initiative on the part of the scout or patrol leader will decide the success or failure of the mission.

PATROLLING IN THE JUNGLE

The commander of troops is more dependent upon his foot reconnaissance agencies in the jungle than in any other type theater. His scouts and patrols will usually be his prime source and often times his only source of information about the enemy and the terrain.

The features common to all jungle areas such as lack of trails, heavy vegetation causing poor visibility for both air and ground observation and the difficulties of cross-country movement by man, animals and vehicles make it mandatory that he send out patrols and also make it essential that he have patrols capable of getting the timely information, which is necessary for him to successfully conduct his plan of operation.

Note

*Prior to the use of this lecture it is assumed that individual training in jungle warfare and living in the jungle has been completed. Relevant enemy tactics and patrolling principles should also have been covered.

The basic principles of patrolling can be applied in the jungle as well as they can be applied in the European theater and the scout or patrol trained to operate in the battlefields of Europe will find that the principles of security, control and movement learned in the European theater must be equally well applied in the jungle. However, to patrol in the jungles the soldier must first know the jungle and its effect upon movement, observation and control. This often presents some new and sometimes baffling problems. It must be remembered that in the jungle theater man fights not only the enemy, but he also fights the jungle vegetation, terrain and climatic conditions. Consequently, he must know how to live in and off the jungle, understand its vegetation and how to cope with the climatic and terrain conditions. After he has learned this, along with basic patrolling principles, the special problems of jungle reconnaissance can best be tackled.

There are three types of jungle: primary, secondary and coastal. Primary jungle is natural vegetation that has never been touched by works of man and has remained in its original state. This natural growth of trees and underbrush is very dense in the valleys and tends to thin out as the elevation increases. Secondary jungle is found where the primary jungle has been originally cleared and then allowed to grow back to a natural state. Here vegetation takes the form of very dense ferns and brambles and any troop or patrol movement is largely confined to exisiting trails. Movement off the trails is very slow as it requires the use of knives and machetes. Along the coastal areas the jungle usually becomes more open and largely consists of mangrove swamps and fields of Kunai grass. Operations in the swamps are naturally slow and difficult. Movement through the Kunai grass is often slow and laborious, as the grass may be anywhere from three to nine feet in height. Fast movement betrays one's presence to enemy air or ground observers who are often located at the edge of jungle surrounding these grassy areas. Broad sluggish rivers are frequent in the valleys and deep fast streams in the hills. Swamps in varying sizes and depth are usually near these rivers and streams.

A patrol starting out on a mission from the coast inland might encounter any or all of the jungle terrain conditions in the course of its journey, depending on the length of that journey. The first hour might be easy going along a smooth coral track running through a coconut grove. Then the patrol might have to work its way slowly through Kunai grass over the heads of the men.

It might have to pass over some of the jagged, slippery, coral ridges that lie across its route and then descend into a valley encountering a section of primary jungle, which would consist of solid walls of undergrowth, thorny vines and creepers. It might take an hour to cut through 300 or 400 yards of this. Next, a patrol might come to a swamp 400 or 500 yards in width and from ankle to knee deep.

It is obvious that aside from any enemy action, the nature of the terrain and vegetation greatly restricts individual patrol activity. In addition, the weather during the patrol journey may have varied from blazing sun to a torrential downpour. The effects of the heat, the perennial dampness, and the rain on the movement of the individual patrol as well as upon the health of the individual soldier are very great and have a decided effect on patrolling operations.

The verdant undergrowth of jungle terrain simplifies individual concealment and the ordinary rules of camouflage apply. Green, yellow, and red paints are good for use on the face and hands and in some cases camouflage jungle suits and packs are useful. However, the ordinary fatigue suit furnishes a reasonable degree of protection without any alterations. It must be remembered that the ease in concealment which the soldier finds because of natural conditions in a jungle is equally applicable to the Japanese, who is a past master in the use of those camouflage materials provided by nature.

As far as the Allies are concerned, night patrolling in the jungle is kept to a minimum. American troops usually pick a bivouac site and establish a perimeter defense before darkness falls. Darkness in the jungle is all that the name implies. It is often impossible to see the proverbial hand in front of the face. Moonlight penetrates jungle growth very little and although you may be able to see the stars and the moon overhead, little illumination seems to penetrate the undergrowth to the ground. Combat experience has shown us that American troops can gain little from night patrolling, although this is a common tactic of the enemy who undertakes his infiltration tactics under cover of darkness.

Because visibility in the jungle is so low, men fighting there must often depend almost entirely upon their senses of hearing and smell. Cultivation of the art of perception must be stressed. Silence and patience are prerequisites of the jungle scout or members of the jungle patrol. Many soldiers are dead today either because they lacked patience, or because they talked when they should have been listening.

Aside from the normal difficulties of moving through jungle terrain, care must be taken by the scout while moving to prevent the vegetation, which he relies upon to conceal him, from also disclosing him to enemy eyes. When operating off trails, it is almost impossible to avoid grasping trees or bushes when working ones way along. However, trunks of slender palm trees and other high trees should be avoided as a means of support while moving and should not be brushed against as the scout makes his way along. The crowns of these thin trees may stick out far above the other vegetation of the jungle and moving the trunk will cause the top or crown to sway indicating movement to enemy eyes, even though the cause of the movement may not be visible.

In addition to the trees, innumerable vines stretch in all directions and care must be taken, when passing through, to lift these vines over the head or step over them without exerting too much pressure, as they in turn will set tree tops or other vines in motion, if too much pull is exerted upon them. In other words, the scout should be careful to grasp or touch only short, sturdy vegetation during his movement.

Although jungle birds are as a rule more sedate than those found in other hemispheres, sudden movement may disturb them and betray the presence of a scout or the patrol.

The following general precepts of movement in jungle terrain should be considered:

(1) The scout will not take cover in places where the overhead movements of vines and trees may betray his presence.

(2) Cover should always be taken where a quick get-away is possible. Mazes of vines and roots provide a definite hindrance.

(3) Streams and water holes should be observed from a distance, if possible, because the presence of natives, animals and the enemy may result in betrayal of position.

(4) Trees as hideouts may be used. However, the danger of a swaying top or branches and the leaving of tell-tale marks on the bark or the breaking off of small branches while climbing may lead to discovery.

(5) Native villages are not always friendly and should be avoided unless the sympathies of the inhabitants are known. For instance, Viet Cong were very often found in villages where they obtained supplies and protection.

(6) The roots of banyan trees, which often times extend high above the ground and form a kind of walk, afford a good place for concealment. Hollow trunks of large uprooted trees are also good hiding places.

(7) Dark shadows near the trunks of trees may indicate the presence of a tree sniper. Traces of climbing irons, sometimes used by enemy scouts, may also be discernible on the lower part of the tree trunk.

(8) Broken spider webs, cut vines and broken branches are signs of the movement of men. Cigarette smoke, cigarette paper, gum wrappers, discarded banana peels, charred wood, human feces, and broken coconuts are other indications of human presence.

(9) A scout should learn to distinguish noises made by man from those common to the jungle, both day and night. Jungle noises such as those made by the movement of the large land crabs, wild pigs, different birds, lizards, surf and rain are often misinterpreted for noises made by the enemy, particularly to the untrained ear.

(10) Foot and boot tracks made by the enemy can be differentiated from those made by your own troops or by local sympathizers. This was especially true of the WW II Japanese jungle troop, who wore a very distinctive rubber-soled shoe. Be aware of similar situations, and be prepared to take advantage of them.

(11) The enemy may be very careless in trail and bivouac security, consequently leaving the above mentioned signs of his presence to be observed by the diligent scout. Oppositely, he may have left boobytraps.

(12) In movement down jungle trails the scout must learn to look up into the trees and also look down his trail for traces of the enemy. It is usually a good idea to have two scouts precede a patrol, one watching the trees and the other the ground.

(13) Friendly natives are a great help in the jungle. Their inherent sense of direction and ability to locate enemy's ambushes is uncanny. They can and have been used as guides, pack carrier and scouts.

(14) A strong knife, an oil compass and the carbine or M-3 sub-machine gun are necessary for successful operation in the jungle.

Because the very nature of the jungle limits observation, movement and control, we can best compare the difficulties of a daytime patrol in the jungle to that which we would encounter in the European theatre over rough terrain at night. Consequently, the time and space factor in planning daytime jungle patrolling operations must be given careful consideration because normal missions, which elsewhere might be designated in terms of miles or thousands of yards in the jungle may be conducted in the terms of "hundreds of yards". This is particularly true in travel off the jungle trails. This factor should always be considered during the planning stage of any patrol operation.

The use of the individual scout as such is not too frequent in jungle warfare, most reconnaissance being done by either the small three or four man sneak patrol or by the heavily armed fighting or combat reconnaissance patrol, usually made up of a squad or more. The smaller sneak patrol is usually forced to stay off jungle trails and natural avenues of approach and must work its way through the undergrowth. Consequently, some of its principal uses have been on missions of limited nature or on those where enemy opposition was not heavy. The larger combat reconnaissance patrol can use dry stream beds, animal or man-made trails to approach its objective. Due to the greater strength in numbers and armament of this type of patrol, it is usually strong enough to fight off enemy attack from ambush, or meeting engagement, by wiping out the opposition, or if too strongly opposed, it can withdraw under protection of its firepower.

Patrol formations in the jungle are usually made up around the basic column formations, which may operate without flanking guards, the heavy undergrowth preventing the flanks from keeping up with the main body, which moves down the trail. The interval in the column is close, usually not over five yards and if traveling at night, physical contact with the man in front, by means of straps or sticks or special markers, may be necessary. In the movement down the trail, one-half of the patrol should be watching the right side, the other the left, and the scout or scouts in front of the patrol maintaining ground and above the ground observation. These scouts usually operate at a distance of no more than thirty yards ahead of the first man of the main body (usually the patrol leader) and should be the most skilled and observant soldiers available.

In an engagement with the enemy, a sneak patrol naturally will fight only in self-defense. On the other hand, the larger combat reconnaissance type patrol is equipped and should be trained to meet the enemy at close quarters. Such a patrol, maintaining a constant state of alertness, prepared for instant action with its weapons, and well-trained in instantaneous employment in outflanking tactics, can successfully cope with the enemy in the jungle.

Because meeting engagements and enemy ambush are encountered at such close quarters, it is advisable to have the source of automatic firepower (BAR, sub-machine gun), near the point of the patrol so that base of fire can be built up, which will enable the rear elements of the patrol to maneuver.

In the event that enemy opposition is too strong, the patrol can withdraw under cover of its automatic firepower or smoke. It is apparent that a well-rehearsed patrol versed in enemy tactics and reactions to ambush and meeting engagements is necessary to be able to perform this type of jungle patrolling operations. (Incorporate in lecture, "Japanese Patrolling and Ambush Tactics." See Foreign Ground Reconnaissance Chapter).

It might be mentioned here that a successful tactic used in areas where the enemy has been particularly active is to send out two patrols along the same trail, one following the other from a one-half mile to a mile. Due to the fact that the enemy may react over-aggressively to our patrols, this particular method of patrolling has worked successfully. The contacting patrol, when encountering enemy opposition, would then withdraw down the trail through the second patrol, which would have set up an ambush upon leaving the first patrol's contact. The enemy following aggressively would be led into the ambush of the second patrol.

One of the best criterion for successful patrolling is to learn from the experience of patrols that have already operated against the enemy. Here is a typical combat-reconnaissance patrol organized from an I-R platoon and dispatched on a four day mission in New Guinea:

It consists of one officer and twenty enlisted men and five native carriers. The patrol is armed with six machineguns and 15 assault rifles, some of which are equipped with grenade launchers. A normal ammunition load plus several grenades each is carried. The natives are unarmed. The uniform is two-piece fatigues with fatigue hat, G.I. shoes and leggings. Each man carries a haversack containing a poncho, one ration, headnet, extra pair of socks, halazone, salt and atabrine tablets. Additional rations are carried by the native carriers. The rations are 'C' rations figured on four cans per man per day instead of the usual six. This is to be supplemented by native fruits and vegetable whenever possible. ("K" rations would have been used had they been available.)

The patrol carries an SCR-610 for communications. (The 610 weighs sixty-five pounds and has a range of about five miles.)

The patrol travels along a main trail and a staggered column is employed. The patrol formation consists of a point, an advance group, the native carriers, the support group, and a rear point. The point consists of two tommy gunners and the patrol leader. The advance group: the first squad of eight men, two carrying Tommy guns. The native carriers next. The support group: six men from the second squad. The rear point: two Tommy gunners and the second-in-command. The distance between men varies from five to eight yards; between groups varies from ten to fifteen yards. Visual contact is always maintained. At times, the natives are placed in rear of the support group.

Each man in the column formation is assigned the same duties that he would have had if a more open formation were possible. For example, the number one point man observes ground to the front, the number two point man observes snipers to the front; the next two men (skipping the patrol leader, who is concentrating on control and the terrain) observes ground and snipers, respectively, to the right; the next two men observe to the left, and so on throughout the squad with the tailing men observing ground and snipers to the rear. A "Get-away" man as such, is not to be designated until trouble developed. (Often the trouble will come from the rear in which case one of the point men may be the logical selection for a get-away mission.)

From time to time the point man halts the patrol for a listening stop. Traveling through the jungle is noisy business, both for you and the enemy. Further, bird and insect noises or lack thereof may well indicate what lies ahead. The rear point drops back frequently for listening stops. (This habit will prevent surprises from the rear.)

At all halts, the patrol moves off the trail and sets up a temporary perimeter of defense, regardless as to whether the halt is for a listening stop, a scheduled rest period, or contact with the enemy. This patrol has scheduled rest halts, of ten minutes duration, every hour. No smoking is permitted. In the stagnant or slow drifting air of swamps and rain forests, cigarette smoke may linger for hours; any passing enemy patrol would be sure to note it. Also, they might note the butt or match. No talking is permitted on any halts except in an emergency and then only in low whispers. (Talking interferes with listening as well as giving the position away.)

On contacting the enemy the patrol plans to move off the trail, each man to his previously assigned and rehearsed position. The patrol leader with the rest of the point is to make a reconnaissance or estimate to determine if the resistance can be driven off or by-passed, always remembering his mission. The second-in-comman is to move up to direct the defense in the absence of the patrol leader on any reconnaissance. In event of an outflanking tactic the second-in-command is to lead the support group in this maneuver.

Locations for bivouac are selected and prepared prior to darkness and are not occupied until nightfall. The bivouac area chosen is usually astride a ridge or a small hill (never bivouac near a stream or on low ground). Groups of three men dig slit trenches on the perimeter; each man knowing the exact location of the other groups.

Hot suppers are prepared away from the bivouac area, and all calls of nature are taken care of there also. All waste and rubbish are buried.

After occuping the position, all men remain awake until one half hour after darkness, and then one in each group of three stays on the alert at all times. No lights or fires are permitted after dark. Necessary movement after dark is prearranged and kept to a minimum. The guards keep in contact with each other by the use of strings, and a very simple system of pull signals is used. All personnel are awakened about an hour before sunrise to remain on the alert in their trenches until dawn. This precautionary measure is to prevent any surprise enemy attack.

Combat-reconnaissance patrols such as the one described above must have a lot of automatic firepower; ideally, at least one automatic weapon for every four men. Other weapons used beside the sub-machine guns, BARs, and assault rifles are grenade launchers, hand grenades (smoke and fragmentation), bazookas and light mortars.

Here are a few points to remember about Jungle Patrols:

1. Travel light.
2. No talking (use only hand signals or prearranged signals or low whispers.)
3. Lots of automatic weapons, concentrated forward.
4. Designate men to observe to flanks, rear, and trees.
5. Designate rendezvous points as the patrol progresses.
6. Thoroughly investigate all possible ambush sites, streams, defiles, etcetera.
7. Have scheduled halts.
8. Set up perimeter defense at all halts.
9. Engage the enemy and move off the trail as soon as contacted. Defensive and offensive plans should be made in advance and rehearsed.
10. Always maintain part of the patrol as a support to exploit a gain or to cover the withdrawal of the advance group.
11. Organize the bivouac area early and occupy it as soon as it gets dark.
12. Patrols should have practice missions and rehearsals in order to function as a team.

Remember, to patrol successfully in the jungle, you must master the jungle, master yourself, and then master the enemy.

CHAPTER VI

EXERCISES

"The realities of battle must be reflected in any training program."

Any manual containing text material on a subject to be used in training should also contain a section describing how this material can best be presented in instruction.

Theoretically, every officer should be a competent instructor, but such is not the case, especially in scouting and patrolling. You cannot rely on the individual initiative of every officer to devise and present training in given subjects to the best advantage. He frequently lacks initiative and creative imagination in presenting a subject in which he has little direct personal interest. Many times he lacks sufficient professional background and necessary instructional data.

The exercises discussed in this section must be adapted to the equipment of individual units and other local training conditions. All that can be done is to suggest what may be accomplished under ideal training conditions. Although some of the following exercises may require preparation and training aids not ordinarily available, all have been designed to fulfill certain needs and to give the soldier considerable field experience in scouting, patrolling, and observation under simulated battle conditions.

Too often, training films have been the major medium of instruction. They are not a substitute for field work and should never be used as such. Training films are used best when they are followed or preceded by classroom instruction and immediate outdoor demonstration and participation in the subject covered. When films are used as the major medium of instruction in scouting and patrolling, the soldier acquires a false estimate of his own ability in fieldcraft, since it looks so easy on the screen. It must be recognized that films such as "The Daylight Reconnaissance Patrol", "Amphibious Reconnaissance Patrol", "Reconnaissance Patrols at Night", or "Reconnaissance Scout", depict the ideal patrol or the ideal scout on ideal missions, and give ideal solutions. The ordinary film includes too many separate points to be remembered unless they are practiced individually in training. Rather than begin a training program in scouting and patrolling with advanced training films, the instructor should thoroughly acquaint his men with the duties of the individual scout so that they will have a better grasp of the instructional points presented on the screen.

A proper mental attitude is very important for men undergoing training in scouting and patrolling. Every effort should be made by the instructor to put realism into all demonstrations and exercises. The American soldier is too accustomed to artificial means of transportation to get any pleasure out of getting down on his belly and crawling. The attitude of "What the hell, this is only a maneuver" is something which must be counteracted from the start. It can be done by several means, all of which require close supervision by the instructor.

The film series "Fighting Men" can be used. Such films of this series as, "Kill or Be Killed", and "Baptism of Fire", will help present the realities of combat. The men being trained should all undergo an infiltration course and if they have not been subjected to live ammunition and demolitions, the training officer should make arrangements for such a course. If there are men present in the area who have recently returned from the fighting front, their personal experiences should be brought before the students. All of the above-mentioned devices will help create an attitude of interest and cooperation during training. Such an attitude is imperative because the nature of the exercises will allow the student with the "lackadaisical" attitude to "let down" when close supervision is impossible during some of the training phases.

Checks and controls (for instruction) must be put into the problem to enable close supervision. For example, a patrol exercise should be used where it can be canalized through certain areas so that the personnel (acting enemy) operating against the students will make more opposition. After study it can usually be determined what checks and controls can be used without violating the tactical aspects of the exercise.

Realism, which is so essential in training troops for combat, must be interjected in training for scouting and patrolling by using tactical problems in which there is actual opposition. *You cannot teach these subjects when the situation and enemy are simulated.* If no friendly troops are available to furnish opposition for the problems the class should be split in half, one group working against the other.

It has been necessary, in some cases, to impose a penalty system for those scouts and patrols who make errors through carelessness or lack of interest. Such penalty systems will gain results when all else fails. The instructor should designate common errors and penalties pertaining to these errors prior to the exercises. The penalty can be extra work, or extra exercises at some future date which will be determined after the course has been finished and an inventory made of the various errors committed by the entire class.

A more direct method can be used if the area for the problems is some distance from the barracks. In this case the instructor can set up penalties which involve the student's walking so many miles back to the barrack area after the problem has been finished. In other words, failure to use cover and concealment could result in the student being let off two miles from his barracks. The seriousness of the offense should regulate the distance to be walked. The student should constantly be reminded that this time he is getting a chance to walk back to camp, but in battle he probably will not. This last method has been used with marked success in providing a needed impetus for classes of both officers and men. Competition fostered between patrols or individuals will also help a great deal to insure a proper attitude toward the training program.

In a theater not all of the unit's time is spent in fighting. The unit must constantly strive to improve its training and proficiency in the preparation before the next battle. Training in scouting and patrolling must be continuous and not regarded complete once the unit takes the field. The battle itself is the best possible training. Full value from combat is obtained when periods of action are followed by intensive training to inculcate the lessons learned. Replacements in particular must be given training in scouting and patrolling.

The chronological exercises presented here are brief and subject to individual change. They should be considered a pattern for the training and although complete enough in most cases, they should be adapted to the special needs of units or individuals in training. Anything to add realism or to include material found in Training Manuals such as 21-75 should be used.

If the training exercises can be done in their chronological order and without outside interferences such as guard duty, fatigue details, and other routine duties, more concentration and benefit will be derived from the program.

The completion of one exercise should not be construed to mean that it is sufficient and no more training is needed on that phase. It should be repeated until proficiency is achieved before going to more advanced problems.

After the basic exercises, the *majority* of the training in *scouting and patrolling should be done at night.* When the student has mastered an exercise in daytime, he should be trained until he can accomplish the same thing at night.

When at all possible, many of the exercises should be worked on *foreign maps* and *photo maps.* This is most important as these mediums are often the only ones available outside of America.

Suggestions and corrections should be made in the field at the time of the error. After the exercise a general critique should be given to the entire group while lessons learned and mistakes made are still fresh in the students' minds.

It will be noted that no time limit has been set for any specific exercise. This is because the size of classes and local training conditions will affect the length of the instructional periods. These exercises should never be hurried. Too often when a definite time is allotted for a given subject the instruction is hurried to present it in that time. In scouting and patrolling, patience and time to perform the mission are basic principles.

Throughout the exercise program a close study should be made of the characteristics of each student as they show up in the problems. Lack of patience, excitability, fear, nervousness, and other traits not desirable in reconnaissance personnel should be noted. Consistent evidence, over a series of problems, of these undesirable traits should result in replacement of the individual concerned.

The first four exercises described here are basic for all reconnaissance personnel. Pictures and a description of the various training aids are included. The scout should master these basic courses before he continues with the other exercises. Violation of the principles of silent movement, camouflage, etc, will often mean failure in his combat mission.

EXERCISE I

PURPOSE:

1. TO TEACH CORRECT CREEPING AND CRAWLING.

2. TO PROVIDE A DAILY PHYSICAL CONDITIONING EXERCISE FOR CREEPING AND CRAWLING.

DESCRIPTION:

A creeping and crawling course should be constructed by making a shored ditch covered by loose barbed wire in three ten yard sections, twelve inches deep for the flat crawl, seventeen inches deep for the side crawl, and twenty-one inches deep for the creep.

The students should be made to crawl from one end to the other. To derive the greatest benefit from the course, the students should carry a weapon with them as they crawl and creep. An instructor should be at each ten yard section for correction purposes.

SUMMARY:

Both the instructor and the wire provide checks for improper creeping or crawling. The instructor also checks for any unnecessary noise. This course should be run daily for conditioning of the students' leg and arm muscles. It can be used in place of morning calisthenics. A course can also be added to this section showing and practicing the proper way to pass through and cut barbed wire.

Prior to participation by the students, techniques and types of crawling should be demonstrated by the instructor with emphasis on where the different types of movement are best used. Daytime use of this course should be followed by night participation.

EXERCISE II

PURPOSE:

1. TO DEMONSTRATE SOUND OBSTACLES THAT THE STUDENTS SHOULD AVOID IN COMBAT AREAS.

2. TO TEACH THE STUDENT TO CROSS UNAVOIDABLE SOUND OBSTACLES WITH THE LEAST POSSIBLE NOISE AND DELAY.

SILENT MOVEMENT COURSE

DESCRIPTION:

The sound obstacles incorporated in the course should be typical examples of terrain. Each section should be approximately ten yards long. A brief, well-worded sign explaining the obstacle and the technique of crossing it should precede each section. Such a course can be made locally in as much as most of the materials used in its construction are readily available. By constant maintenance and variations in the size of certain sections large units can utilize this training aid. As the student walks through this course he must maintain his balance and be able to shift his weight from his rear foot to the forward one after he has placed it silently on the terrain. Heel first, toe first, edge of foot, or flat foot technique may be used when the forward foot is placed on the ground. Any method by which the individuals can move silently should be permitted.

SECTION I
DEBRIS # TILE

1. This section should contain typical rubble and debris. It will give the students practice in balance and patience. The student should go through the debris itself without trying to skirt the edges.

2. The proper procedure is to test the debris with the hand, remove anything that will break, then move forward.

Figure 72

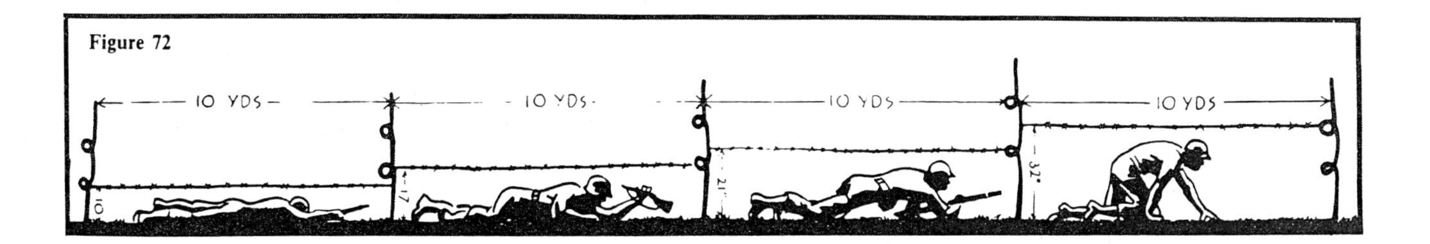

SECTION II
TALL GRASS

1. This section should include grass and weeds at least twelve inches high.

2. In long grass lift the foot high and bring it down flat. Here it is most diffic·.lt to maintain proper balance as well as silence.

SECTION III
TURF WITH OCCASIONAL OBSTACLES

1. Typical turf with an occasional twig, leaf, or branch.

2. To cross this kind of terrain the student has only to look before he steps.

SECTION IV
MUD AND MUCK

1. This area should be filled with soft, watery mud. It is almost impossible to avoid a sucking sound when the foot is lifted. Such a sound trap should be avoided, but it can be crossed silently, if the shoes are wrapped with burlap rags, or if snow shoe like attachments improvised of twigs are attached.

SECTION V
LOOSE ROCKS

1. This section should have about ten yards of loose rocks of varying sizes and shapes.

2. The proper way to cross rocks is to test each rock before stepping. It is better to step with the foot flat in order to keep balance.

SECTION VI
SAND

1. Ten yards of loose beach sand to form a track trap.

2. The sand is noiseless, but whoever walks in it will leave tracks. These tracks must be smoothed out with the hands or leaves, etc. Note: casts of American and enemy boot tracks, tire marks, etcetera, can be displayed on each side of this section.

SECTION VII
TURF WITH OCCASIONAL OBSTACLES

1. Same as Section III.

2. This section is repeated because this kind of terrain is more common to all countries than any other type. The types of obstacles may be changed.

SECTION VIII
GRAVEL

1. Two-thirds of the width should be laid on a wood foundation. Along the sides the gravel should be laid on the ground. This stimulates a gravel road.

2. The student will find that he can move more quietly and quickly on the sides where the gravel has become more packed down, as in wheel tracks. On the wood, in the center, the pebbles remain dry and separated causing a grating sound when stepped on. The most satisfactory method of crossing gravel is to use the foot flat.

3. The use of wood merely emphasizes the grating noise. Danger of mines on the shoulders of roads should be mentioned by the instructor when students are passing through this section.

SECTION IX
DRY LEAVES AND STICKS

1. Ten yards of brittle sticks and dry leaves.

2. This is a sound trap to avoid in combat. To cross successfully use the hands to test and remove the twigs before stepping forward. The enemy will always listen for the crackling of undergrowth, because it is his chief guide to hostile presence.

SECTION X
LOOSE STONE FENCE

1. A stone fence 3 or 4 feet high and 3 feet wide with loose rocks on top. (common in European Theater).

2. To cross the stone fence the student must keep the silhouette small and be careful not to brush the loose stones. He must lower himself silently to the other side.

Speed of completion should be considered. Anyone can be silent when he moves at a snail's pace.. The proper speed is a medium-slow pace. Over-caution may be a man's downfall, especially if his sense of balance is bad.

Balance is the most important single consideration in crossing all obstacles. To keep the balance, bend the body so that there is as much weight behind the feet as there is in front of them. When the student moves, his weight should be over the foot on which he has landed. He must learn to transfer his weight as soon as he has satisfied himself that his forward foot is not going to break a twig or some other object which might reveal his positon. Then, while he is waiting to take the next step, he should distribute his weight evenly on both feet.

The silent movement course is especially valuable for reconnaissance scouts and patrols. It helps develop confidence in the man's ability to move silently and teaches him that sound obstacles can be conquered. *It should be used repeatedly in training and should be used at night as much as day.*

Figure 73

Figure 74

Figure 75

Figure 76

Figure 77

Figure 78

Figure 79

Figure 81

Figure 80

Figure 82

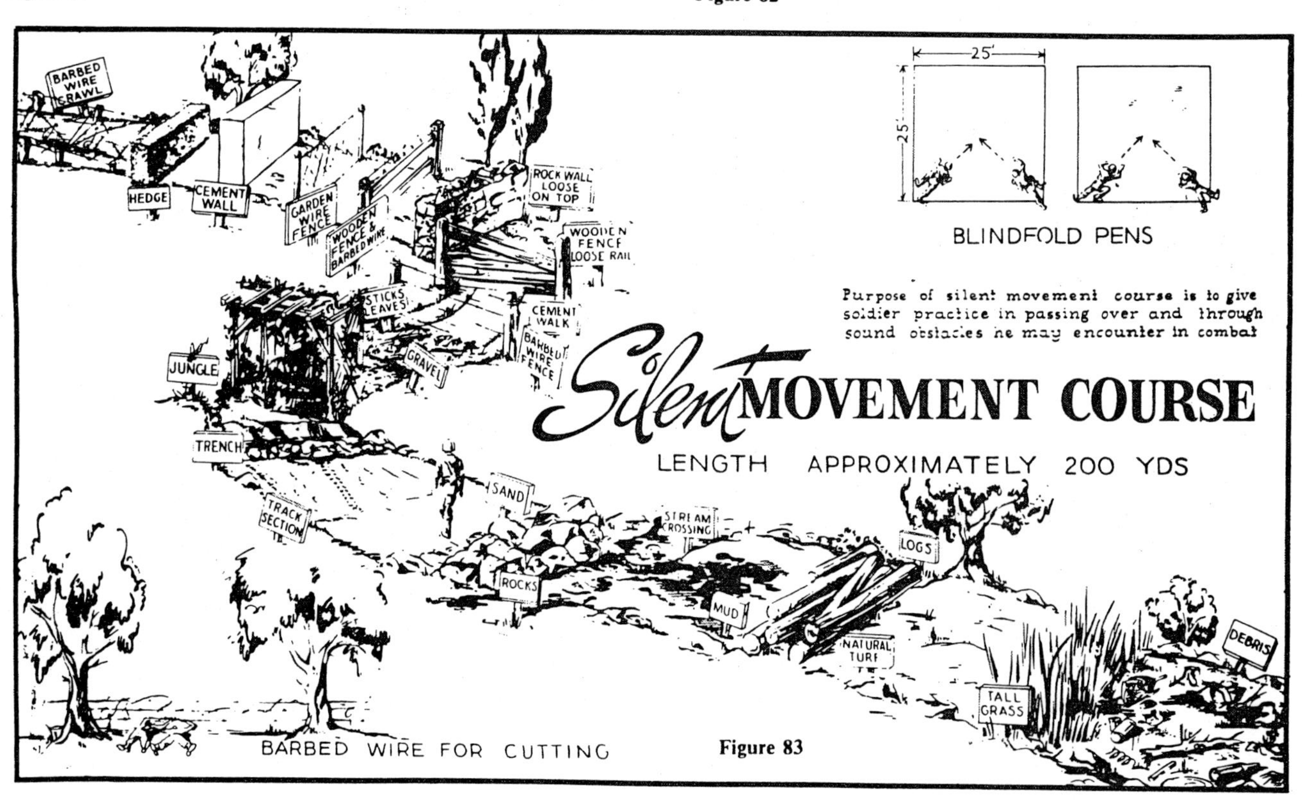

Figure 83

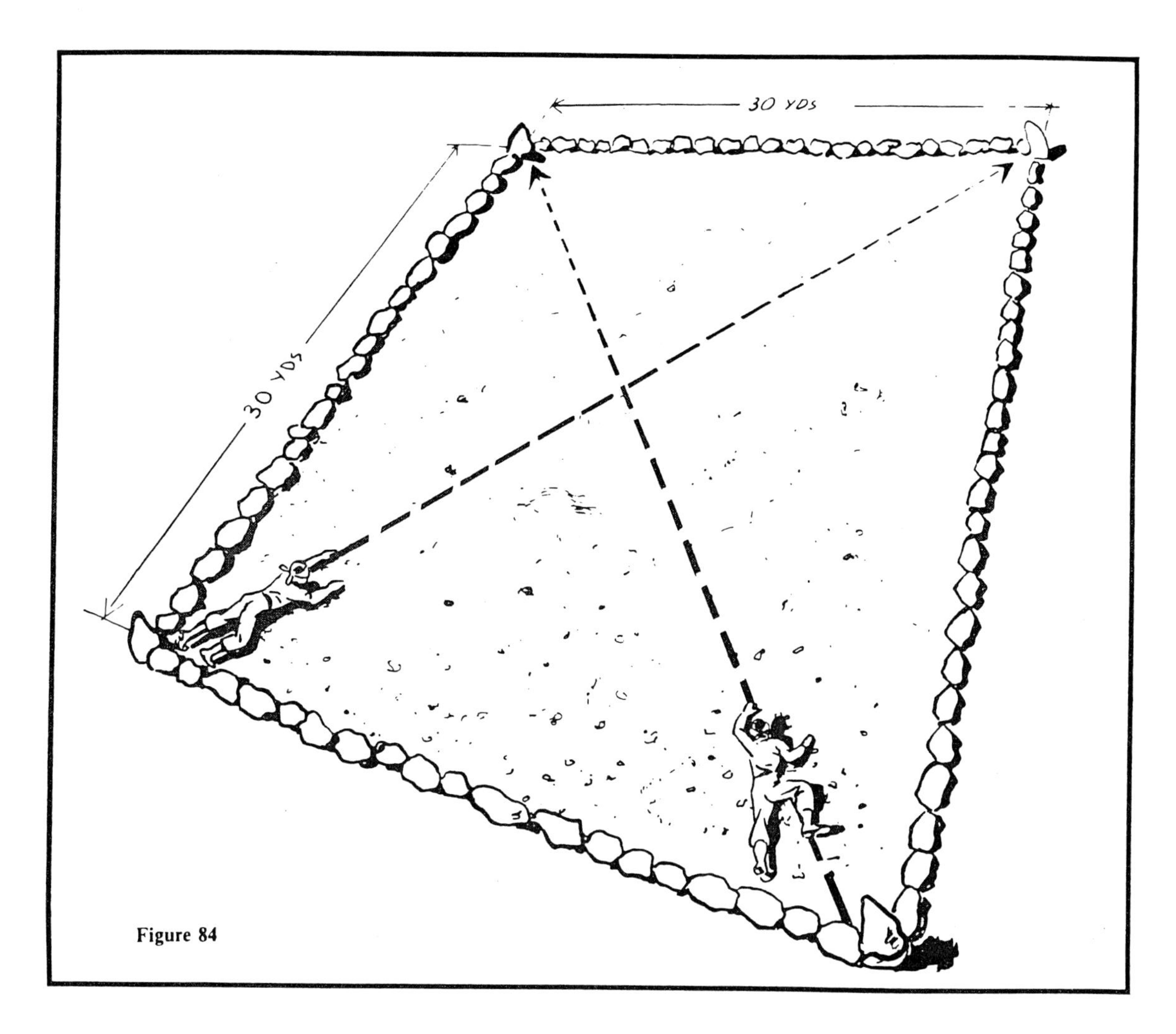

Figure 84

It must be emphasized in all the instruction that although the purpose of the course is to teach proper silent movement over the various sound obstacles, they should be avoided in combat whenever possible. Knowing what they are, the scout will avoid them on his mission if he possibly can.

Additional sections such as piles of logs, trenches, pools of water and dense undergrowth can be added to the course as desired.

EXERCISE III

PURPOSE:

1. TO DEMONSTRATE THE IMPORTANCE OF AVOIDANCE OF SOUND IN NIGHT MOVEMENT.

2. TO CHECK SENSE OF DIRECTION AND THE ABILITY TO CRAWL SILENTLY.

3. TO EMPHASIZE THE NEED FOR CAREFUL DAYTIME TERRAIN STUDY PRIOR TO NIGHT MISSIONS.

DESCRIPTION:

Build an enclosure about 30 ft. square, using low fences of logs or rocks for the boundaries. The ground in the enclosure should have some minor variations. (Crawling pens.)

Place two students in corners on the same side and instruct them to study the ground thoroughly inside the enclosure. Then blindfold them and have them crawl silently from their corner diagonally across the enclosure to the other corner. They should avoid contact with each other when crawling across the area.

SUMMARY:

Although the exercise is not an exact duplication of night work, it will show the problems of maintaining direction by sound and feel alone. Many students will end up in the wrong corner and should be made to repeat the exercise.

Noise producing obstacles such as twigs can be added to make silent crawling more difficult. Emphasize to the students how a study of minor irregularities in the terrain will help them orient themselves when crawling to the diagonal corner.

EXERCISE IV

PURPOSE:

1. TO DEMONSTRATE INDIVIDUAL CAMOUFLAGE.

2. TO SHOW THE STUDENTS HOW EVEN THE SLIGHTEST MOVEMENT MAY DISCLOSE A CAREFULLY CAMOUFLAGED POSITION.

DESCRIPTION:

After a lecture, movies, and demonstration on camouflage, the students should see an actual demonstration of camouflage. In a small area, usually on the side of a hill which provides some natural cover, a squad of men should be carefully camouflaged so that very few of them can be picked up by the observing students. The nearest position should be not more than 10 yards from the students, but this position should be the most carefully chosen and camouflaged. The concealed men should be numbered by the instructor for better control.

After the students have arrived and an introduction has been given, the instructor giving the limits of the area should challenge the students to pick out the hidden men. After a few minutes the instructor should ask for volunteers to point out any of the hidden soldiers. As each student picks out a hidden man, he should reveal the position to the instructor only, thereby giving the less observant students a chance to benefit more from the exercise. The instructor signaling the remaining men by their number should instruct them to make some slight movement such as scratching the head, picking the nose, or stretching, to show the students how these simple movements can give away the most elaborately camouflaged positions. As a finale, the closest undiscovered man can fire a blank or throw a fire-cracker at the class.

SUMMARY:

The demonstration should do two things:

(1) It should prove to the students that a skilled soldier can easily conceal himself a short distance from the most observant eyes.

(2) It should also prove that even the simplest movement can betray a position. This exercise can be elaborated on by a demonstration of the proper use of various artificial camouflage aids such as nets, paint, and camouflaged suits.

TF 5-645, "Camouflage, Individual Concealment", should be shown and demonstration of the principles depicted should be made. TF 7-234, "Use of Natural Cover and Concealment", can also be shown.

EXERCISE V

PURPOSE:

1. TO SHOW THE CAPABILITIES OF A SNIPER.

2. TO DEMONSTRATE THE NEED FOR EXTREME CARE AND UTILIZATION OF COVER IN HOSTILE AREAS.

DESCRIPTION:

A sniper armed with a rifle equipped with a telescopic sight, should be hidden in a well-concealed position. On the bank there should be a number of small easily broken targets such as electric bulbs, old saucers, etc. Near these smaller targets the standard American helmet should be placed for effective contrast. At prearranged signals the sniper fires, breaking the targets.

The students, with due regard for the safety factor, should be placed close to the targets on the bank. It is not necessary that the students see the sniper's position, but they must be close enough to get the full effect of the sniper's bullet and accuracy. The instructor should bring out the obvious contrast between the size of the targets and the helmet. He should also explain how a sniper operates and his capabilities. It should be emphasized how they usually cover trails, crossroads, and other places of forced passage and how they strive to pick off officers, patrol leaders, and other key men.

SUMMARY:

This exercise will vividly emphasize the need for skill in the use of cover, concealment, and movement. After the demonstration each student should look through the telescopic sight so that he can more readily understand how the scope adds to the sniper's effectiveness and powers of observation.

It is advisable to precede this exercise by a lecture on the combat use of the sniper.

EXERCISE VI

PURPOSE:

TO ACQUIRE PROFICIENCY IN RANGE ESTIMATION.

DESCRIPTION:

After a preliminary lecture on principles of range estimation, have the students practice by estimating the range to objects at previously determined distances. Start out with short distance estimation, gradually increasing the range as proficiency increases.

Explain and have the class practice the use of the mil scale in binoculars for range estimation. If available show and explain the use of an aiming circle and a range finder.

Show and practice the estimation of ranges by sound, using prepared demolition charges at long ranges.

SUMMARY:

Skill in range estimation by use of the eye alone cannot be accomplished in one instructional session. It must be practiced constantly in all terrain and light conditions. If five minutes a day can be spent in practice at estimation of distance to objects of different sizes and shapes, real skill can ultimately be developed.

It is desirable to use a range finder prior to the class in order to accurately establish ranges to the objects used for instruction.

EXERCISE VII

PURPOSE:

1. TO DEMONSTRATE THE DIFFICULTIES AND DANGERS OF NIGHT RECONNAISSANCE.

2. TO TEACH THE STUDENT THE BEST WAY TO USE HIS EYES AND EARS.

DESCRIPTION:

The following demonstrations are suggested as a night exercise. A lecture on the proper use of the eyes and ears should be given previously.

I Night Sight Demonstration.

1. Have the students pick out a very faint star and while looking at it instruct them to try to spot a much fainter one out of the side of the eye. When direct vision is shifted to the fainter star, it disappears and come back only when vision is refocussed on the original star.
2. (a) Have a man silently approach the students out of the dark and halt when seen.
   (b) Have the man return into the dark while the students test their ability to follow him for a longer distance once the vision is fixed.

II. Night Sight Range Estimation:

1. Place men with flashlights, matches, lighters, and cigarettes at ranges from 100 to 1000 yards and have these men perform individually to give the students an opportunity to see the amount of light made at these various ranges.
2. For the second part have selected men at various known ranges light cigarettes. The student should estimate the ranges.

The area must be well chosen for this demonstration. It should be a long, gentle-sloping hill, with the students seated at the top of the hill.

III. Night Blindness Demonstration

1. Explain to the student the proper actions when a flare explodes.
2. Have a flare unexpectedly shot off and observe the students' actions. Let them see how a patrol should act, using a demonstration patrol for instruction.
3. Have students look at the bright light of a flare to find out how long it takes to regain full vision.

IV. 1. Place men at intervals, have them talk, dig fox holes, rattle mess kits, pitch tents, cut trees, work rifle bolts, fire blanks, light matches, etcetera.

2. Have the students identify the sounds and flashes, estimate the ranges, take azimuths.
3. Show how the azimuths taken by sound can be used to locate positions by intersection.

V. Sound - Flash Demonstration

1. Place students at least 200 yards from the front of a line of enemy and American weapons. Have them lay prone.
4. Using a radio or phone the instructor should call on the various weapons to fire. Differences in rate of fire, muzzle blast and sound of comparable enemy and American weapons can then be shown.
3. Emphasize the importance of weapon identification by these means in day and night reconnaissance.

SUMMARY:

The instructor should stress the following points:

1. When a patrol hears a flare leaving the projector, the men should fall flat on the ground keeping their faces down.
2. If a flare bursts and catches the patrol standing, the men should remain motionless. Qualify this as to terrain.
3. Members of a night patrol should never look at bright lights as it decreases the efficiency of night observation.

EXERCISE VIII

PURPOSE:

TO GIVE STUDENTS PRACTICE IN SILENT MOVEMENT AND EXPERIENCE IN APPROACHING AND CONTACTING SENTRIES.

DESCRIPTION: (night)

Place several students in full field equipment along a road and have them walk post as sentries. Assign the remainder of the class the job of ambushing the guards, using small bags of flour instead of actual physical contact. The exercises must be controlled at all times, with close supervision in the area where the sentry is contacted. Continue this exercise until every student has had the opportunity to act both as a sentry and later as an attacker of the sentry.

SUMMARY:

This exercise should continue over a period of several hours and should not be hurried. It is important that all students be assigned the sentry duty as well as the ambush, since it is only in this way that they can absorb sentry psychology and understand more thoroughly the problems of stalking. Students should be made to approach the edge of the road from a distance forcing the use of concealment and silence.

A simpler version of this exercise is to have one student blindfolded and have another creep up and try to touch him. If the blindfolded student claps his hands before he is touched, he wins.

EXERCISE IX

PURPOSE:

TO DEMONSTRATE AND PRACTICE THE PRINCIPLES OF SILENT MOVEMENT, PATIENCE, AND CAMOUFLAGE.

DESCRITION: (day)

Select an open field containing high grass or ferns. A few bushes and trees will not spoil the area if it is open enough for clear, all-around observation.

In the center of the field place some numbered white flags. In the area of the flags place some standing guards (no closer than 50 ft. to any one flag.)

Each student should be instructed to approach the field, enter it and pull down his assigned flag without being seen by the guards. Failure should mean repeating the exercise.

SUMMARY:

Failure to select the proper approach, too swift movement causing the grass to move unnaturally, or a lack of patience will be demonstrated here.

If the class is large, half of the students may act as guards while the other half participates. In any event, the students not participating should be watching for the mistakes of the others. This exercise may be done at night if the moon is bright enough.

The area selected and the cover should be such that it is possible to bring down a flag if proper routes and cover are selected by students.

EXERCISE X

PURPOSE:

1. TO TEACH THE NECESSITY OF PATIENCE AND SILENT MOVEMENT IN NIGHT RECONNAISSANCE.

2. TO GIVE EXPERIENCE IN PASSING SENTRIES AND ENEMY OUTPOSTS.

DESCRIPTION: (night)

The area needed should be about 200 yards wide, 500 yards long, and relatively free from vegetation.

At two-thirds of the length, a moving guard patrols the width; near the end a standing guard is concealed. The students must pass both guards. Those who make too much noise or are seen by the sentries are sent back for another try. It is advisable for the instructors to act as guards.

SUMMARY:

As the student approaches the guards, he should use the skyline to pick them out and to pace and study the habits of the moving guard. On a bright night the creep is used to keep down silhouette and shadows. On a dark night the bear crawl can be used. Again the area selected should be such that skillful use of available cover will enable the student to complete it successfully.

EXERCISE XI

PURPOSE:

PRACTICE IN SILENT MOVEMENT AND CONCEALMENT.

DESCRIPTION: (dry)

This exercise can be conducted in any area approximately 200 by 300 yards. Grass, brush, and small trees sufficient to give concealment should be present.

A small group of students should be placed at each end of the field. They should be armed with rifles and blank ammunition. Each group should advance toward the center of the field stalking the members of the other group. Full advantage should be taken of natural cover and concealment.

SUMMARY:

There is great necessity here for close supervision as the problem should not be allowed to degenerate into a fire fight. Instructors or umpires should occupy commanding positions near the center of the field where they can supervise the action and make decisions on casualties. Boundaries of the combat area must be clearly defined.

This type of exercise injects the personal element and can be quite instructive. During waiting periods for other exercises, it can be used between individuals as well as groups.

EXERCISE XII

PURPOSE:

TO DEMONSTRATE AND PRACTICE PATIENCE AND CAREFUL MOVEMENT.

DESCRIPTION: (night)

Place two students in opposite ends of a large, dark room (barn, barracks). They are to advance toward each other. Each student is to stalk the other. Contact should be made by flour sacks, or water pistols. At the moment of contact lights should be turned on.

SUMMARY:

One of the results of this exercise is to demonstrate inability of men to move swiftly and silently in a dark, enclosed area. It should also emphasize how such simple actions as heavy breathing, scraping, creaking boards, and heavy steps may cause failure. This type of exercise requires close supervision and careful control.

Emphasize that the individual who moves slowly and stops frequently to listen for the opponent will ordinarily win.

EXERCISE XIII

PURPOSE:

TO TEST AND DEVELOP THE STUDENT'S MEMORY AND POWERS OF OBSERVATION.

DESCRIPTION:

The instructor will place a few students in a vehicle and proceed at a normal rate along a road. From time to time the instructor will ask questions:

How many people did we just pass?

Were they men or women?

How were they dressed?

We just passed a building. What was it?

etc.

SUMMARY:

From this type of problem the student soon learns the value of being alert and mentally recording all small bits of seemingly unimportant information. This type of exercise with variations should be repeated until good observation becomes a habit.

A number of articles laid on a table that is enclosed by a curtain can also be used to test observation. Pull the curtain exposing the items for a short time then close the curtains and ask the student to write down what he saw, how the items were arranged and other pertinent questions.

It is assumed at this point that instruction has been completed in compass, map reading, terrain analysis, message writing, enemy identification, and other pertinent subjects mentioned in Chapter I of this text.

EXERCISE XIV

PURPOSE:

TO GIVE THE STUDENTS EXPERIENCE IN THE MECHANICS OF LOCATING ENEMY ACTIVITY FROM AN OBSERVATION POST.

DESCRIPTION:

From two or three previously selected observation posts the students should observe prepared enemy demonstrations or incidents to their front and flanks. After each demonstration the students will plot the approximate positions on the map. At the same time they will estimate the range and record the aximuths to each incident and plot it on a ground observer's report sheet.

Messages will be written for each incident and will be graded later for completeness and accuracy.

SUMMARY:

At the end of the problem the instructor will compare the accuracy of the observation post estimations with the correct ones. This will demonstrate to the students the value and need of intersection in locating enemy positions. A discussion of the use of intersection by artillery, air, and mortar personnel should be given. Range to the prepared incidents should also be previously determined to check accuracy of student range estimation.

This exercise should be run with variations until the whole group is skilled in all phases. Foreign maps and photo maps can be used.

Prior to this exercise the students should have been instructed in the proper use of field glasses and other means of range estimation. This exercise demonstrates the practical applications of range estimation, intersection, compass, identification, etc. Tactical OP operation will be covered in a later exercise.

NOTE

If actual enemy equipment, tactics and disposition are used, a similar exercise should have been run previous to this to test recognition of American equipment.

EXERCISE XV

PURPOSE:

TO GIVE PRACTICAL EXPERIENCE IN LOCATING OBJECTIVES AT NIGHT BY SOUND OR FLASH.

DESCRIPTION: (night)

Two listening posts should be set up for the use of students. During the demonstration various weapons should be fired and other sounds made at different distances. The students should estimate the range, identify the weapons or sounds, and determine the aximuths.

After the demonstration, the students should plot the exact position of the weapons from the intersecting azimuths of two listening posts.

SUMMARY:

This exercise will demonstrate the necessity of the listening post as a basis of night operation. The value of muzzle flash and sound as a locator and a means of weapon identification can also be demonstrated. It should be emphasized that the listening post can also be the day observation post and that special listening posts in close proximity to the enemy are also used for this purpose.

Difficulty will be experienced in this exercise by the students if they have not had a chance to study sounds and weapon-flash previously. This exercise will need repeating many times. While the phase of identification or location is important it is advisable to cover it but once and go on through the other exercises before spending too much of limited instructional time on it.

FB - 31, "Battlefield Sounds", can be used prior to the exercise. It must be recognized that first the student must know the sounds of our own weapons and then should be schooled in enemy weapon identification by sound and flash.

EXERCISE XVI

PURPOSE:

THIS EXERCISE IS TO TEACH THE STUDENT THE TECHNIQUE OF COMPASS OFF-SETS IN THE FIELD.

DESCRIPTION:

The exercise must be designed to give each student practice in the work. Following are two methods found most satisfactory.

1. The student is given an azimuth which directs him to his objective. On the way he runs into prepared ambushes, gassed areas, etcetera, which he must by-pass. The student at these points must take compass off-sets to keep from getting lost.
2. The following method has also proven particularly successful in practice. In this exercise the student is given no map, but will have compass, protractor, pencil and paper. He is given the azimuth and the distance to his objective, but he is instructed not to travel across country. The practice is to travel by road and reach the objective without the aid of a map. To do this, he must plot the route carefully to insure his arrival at the proper destination.

SUMMARY:

The exercise will fill in gaps in the individual training of the soldier in compass work. The first method is a good basic problem; the second, an excellent introduction to any mounted off-set problem. The combat use of the compass off-set to by-pass any hostile position or area, when following an azimuth on a mission, should be stressed.

EXERCISE XVII

PURPOSE:

1. TO TEACH THE IMPORTANCE OF SELECTING GOOD OP POSITIONS.
2. TO DEMONSTRATE A GOOD ROUTE FOR A DAYLIGHT RECONNAISSANCE PATROL.
3. TO TEACH THE IMPORTANCE OF GATHERING ACCURATE INFORMATION ABOUT STREAMS, BRIDGE CAPACITY, ETC.

4. TO DEMONSTRATE TERRAIN AND ITS RELATIONSHIP TO THE SELECTION OF AN AMBUSH.

DESCRIPTION:

The students are divided into small groups, and then taken on a tour of five possible OP locations. At each of these locations, the advantages and disadvantages are discussed. After each student has made his decision as to which OP he favors, and given his reasons, the class is given the correct solution and assembled at the location of the best OP. At this location, they select the best route for a daylight reconnaissance to a designated point. This is turned into the instructor. These routes are discussed, and then the group is lead along the route that has been proved the best.

Along the route, the attention of the patrol is directed to: (1) speed of creeks; (2) capacity of bridges; (3) width of streams; (4) possible ambush location.

Lectures on patrolling, ambushes, and terrain analysis in respect to the employment of weapons should have preceded this exercise. In this exercise the instructor may also discuss the differences in selection of route between day and night patrolling, pointing out the differences in utilization of care, proper movement, etcetera, for day and night patrols.

EXERCISE XVIII

PURPOSE:

TO TEST THE INDIVIDUAL'S ABILITY IN MAINTAINING DIRECTION AND SELECTING PROPER ROUTES, OBSERVATION, AND THE USE OF COVER AND CONCEALMENT.

DESCRIPTION: (night - day)

Each student is equipped with a compass and a map of the area. On a map are marked the points where the student detrucks and the point to which he must proceed to reconnoiter and observe. The students should be detrucked individually along a road running in the edge of a wood. The student, then "on his own", proceeds to his objective and observes any enemy activity and installations. At the various objectives, there should be some sort of enemy installation or activity such as machine guns, entrenchments, barbed wire, etcetera. To add realism to the exercise, enemy riflement with blanks may be scattered over the entire area where the students will operate.

After completing his reconnaissance and observation, each student will return to a designated assembly area and report to the instructor.

SUMMARY:

Care must be taken to choose a large enough area so that the students will not "bunch up". Also, the detrucking interval should be great enough so that each man will have a different route. This exercise should be done in daylight and at night. This individual problem is the only way the knowledge of the individual man can be faithfully tested. Show TF 7-280 "The Reconnaissance Scout" prior to this exercise and hold a critique immediately after.

EXERCISE XIX

PURPOSE:

TO DEMONSTRATE PROPER PATROL FORMATION AND PROCEDURE.

DESCRIPTION:

Using a demonstration unit, demonstrate the diamond, its T variations, and the single file formations for reconnaissance and fighting patrols.

The instructor should comment upon and have the patrol demonstrate the following point:

1. *Duties of a patrol leader*
2. *Role of scout*
3. *Intervals*
4. *Control*
5. *"T" or wedge variations*
6. *Proper methods of corssing ridges, streams, defiles, movement by bounds, etc.*
7. *Column and diamond formation*
8. *Road Patrol: staggered or dispersed.*
9. *Security in movement and at halts.*

SUMMARY:

Stress again the role of the scout and the fact that the patrol leader is the head and heart of the patrol.

The diamond and column formation are the basis for all patrols, regardless of size. Emphasize the advantages of the three man "sneak" patrol over larger (5 and 6 man) reconnaissance (sneak) patrols. Stress the use of the Combat-Reconnaissance type patrol as a means of getting information unavailable to sneak reconnaissance.

EXERCISE XX

PURPOSE:

TO GIVE THE STUDENTS PRACTICAL WORK IN THE MECHANICS OF MOVEMENT OF SNEAK PATROLS.

DESCRIPTION: (day and night)

An area is chosen which contains several small hills with some wooded portions. The patrol is grouped behind one of the hills, and a conference is held covering the subject matter relative to sneak patrols. The patrol leader is then given the mission of reconnoitering one of the nearby hills. Enemy activity should be prearranged so that the patrol will have something to observe. Enemy ambush parties should be posted along all possible patrol routes.

SUMMARY:

TF 7-1061 can be shown prior to this exercise. A discussion of the points in the film should be made. This exercise must be repeated day and night until the instructor is satisfied as to the proficiency of the students on this type mission. It is advisable to have an instructor umpire with each patrol to correct errors at the time of occurrence.

A critique should be held where the patrol can view the terrain over which it has just crossed. Points covered in the critique should include:

1. Route chosen by patrol leader, use of cover, etcetera.
2. Patrol formation, control, security, movement.
3. Action of patrol when ambushed.

EXERCISE XXI

PURPOSE:

TO GIVE THE STUDENTS PRACTICAL WORK IN THE SELECTION, ORGANIZATION AND EMPLOYMENT OF AN AMBUSH.

DESCRIPTION:

The terrain selected for this exercise should be wooded with several trails running through the woods. A situation can be worked up suitable to the terrain selected. The students are divided into patrols (ambush parties), briefed on their missions, and issued weapons. The mission of each patrol is to move down one of the trails approximately 300-400 yards (depending on terrain) to set up an ambush for enemy patrols. The patrol leaders and second-in-commands move down the assigned trails to select the ambush points while the remainder of the students wait at an assembly area. When the ambush point has been chosen, the second-in-comman returns to lead the remainder of the patrol to the ambush site. Then by placing each man in position, the patrol leader completes his ambush under the supervision of an instructor. The instructor radios the "enemy" patrols to move out. When the "enemy" patrol arrives at the ambush site, the patrol leader gives the signal for the ambush to open fire. The instructor concludes the exercise with a critique.

SUMMARY:

Prior to this exercise the students should be given a lecture on the various types of ambushes and the employment of weapons in ambushes.

The "enemy" patrols must be rehearsed prior to the exercise so as to be sure they will act normally to the ambush. The use of blank ammunition is essential but supervision must be exercised to prevent a degeneration into an uncontrolled fire-fight.

The difficulties of night ambush, silent ambush, and the importance of proper and prior planning and rehearsal should be brought out.

EXERCISE XXII

PURPOSE:

TO TRAIN THE PATROL IN THE PROPER METHODS OF COMBATTING THE AMBUSH.

DESCRIPTION:

In the same area selected for Exercise XXI, direct a patrol into a prepared ambush by assignment of an azimuth or a prescribed route. During the exercise, the instructor will check action of the patrol scout in searching the danger areas, retention of formations, and other security measures of the patrol.

At the time of the ambush there must be close supervision of all the action. Extra instructors, who will act as umpires, will decide on casualties, proper action of the ambushed patrol, regrouping of formations, rear guard actions, and withdrawals.

Again caution must be observed in not allowing a fire-fight to materialize. Any means such as colored arm bands or large numbers to distinguish between the patrol and the ambush party should be employed.

SUMMARY:

This exercise can be increased in realism by the use of smoke, CN, night, bad terrain, or severe weather conditions. Many variations of this basic exercise are possible to the instructors. It should be pointed out that the best protection a patrol can have against the ambush is proper use of formations and its scout.

The difference between the reaction of a sneak patrol and that of a fighting patrol to an "enemy" ambush should be covered.

EXERCISE XXIII

PROPER PREPARATION AND BRIEFING OF PATROL FOR NIGHT PATROLLING

(Do Not Omit in Any Program)

PURPOSE: TO TEACH STUDENTS:

1. TO ANALYZE TERRAIN IN
TO BRIEF A PATROL PROPERLY AND THOROUGHLY
3. TO EXECUTE A NIGHT PATROL MISSION.
4. TO INTERROGATE A PATROL RETURNING FROM A MISSION.

DESCRIPTION:

The instructor, acting as Battalion S-2, briefs the patrol leaders covering points relevant to:

1. Mission.
2. Routes to be followed.
3. Enemy dispositions.
4. Location and activities of friendly troops.
5. Outpost and other security elements through which the patrol is to pass.
6. Terrain conditions.
7. Missions and routes of other patrols.
8. Time patrol is to return.
9. Place where messages are to be sent, or the patrol is to report.
10. The challenge, password, and reply.
11. The size, composition, and weapons of the patrols.
12. Any special instructions.

After receiving the brief from the Bn S-2, the patrol leaders brief and prepare their own patrols. The patrols are furnished with field glasses, appropriate weapons, and maps of the area, and personal equipment.

After the patrol leaders have been briefed by the Bn S-2, they issue a warning order to ready the members of the patrol. Then they spend as much time as possible in observation of the area to be reconnoitered, so that a most appropriate route may be chosen. After this, they return to the Bn CP and hold a rehearsal. The final order is then issued, and the patrol passes through the outpost to do its reconnaissance. At a predetermined time, the situation is declared non-tactical and a critique of the entire action is delivered by the officer umpire who has accompanied the patrol.

The importance of a proper interrogation of the patrols should be emphasized.

This exercise should be run with variations wherein the individual student will act as the dispatching (briefing) officer who is sending out a patrol or patrols on a mission.

NOTE

Primary considerations for the following exercises are the selection of terrain and careful preparation of the problem itself. Realism should be stressed by having the tactical situations exist from the moment the students leave the initial point until the time that they return. The students may carry all equipment required and live on field rations and field conditions for the duration of the problem. The enemy must be represented by actual troops who will act as any normal enemy would in similar situations.

The longer the problems can be run in the field the more real they will be. Short problems are made unreal by the fact that the "enemy" is on the alert for the short time of the problem and picks up the students more easily. They do not afford the time for adequate observation, proper briefing and planning of patrols, or for actual patrolling.

In order to control the problem and evaluate the work of the students, the assigning of assistant instructors to each patrol has been found satisfactory.

If instruction in aerial photography and foreign maps has been given, photo maps and pertinent foreign maps can be used during the problems.

EXERCISE XXIV

OBSERVATION POST EXERCISE

PURPOSE:

(1) To reconnoiter and select an observation post.

(2) To establish and maintain communication between the observation and the advance command post.

(3) To organize the observation post so as to insure efficient operation and to demonstrate zones of responsibility in observation.

TIME:

1800-1000 the following morning.

DESCRIPTION:

The students are divided into various OP parties with a leader designated for each. The OP leaders and one NCO from each party arrive at the CP before the remainder of the students. There they are given the situation and requirements. Both friendly and enemy situations can be worked up to suit the terrain. The requirements presented each leader are:

(1) To reconnoiter and select an observation post from which his zone of responsibility can be observed.

(2) To establish and maintain telephone communication between the observation post and the command post (sound power phones preferable).

(3) To organize and dig in the observation post, including provisions for relief of personnel, prone shelters, camouflage.

After the OP leaders and their NCO's are oriented they move out and reconnoiter for OP positions. The NCO's return to the CP to guide remainder of OP parties who will lay wire from the CP to their OP's. The OP leader remains at the OP and pinpoints his position on the map, makes a range card, plans for camouflage, traffic control, prone shelters, relief of personnel, roster of men to be on duty, and selects alternate OP site.

The remainder of the OP personnel lay the wire from the CP to the OP. One man remains on duty during the night testing his communications every 30 minutes. If breaks occur (these may be planned), proper repairs of the wire must be made by the students.

At dawn enemy activity occurs for two hours. Ground Observer Reports and situation overlays will be kept at each OP. Messages will be written correctly before being telephoned to the CP where the information will be plotted on a large map. For practice, activity in odd numbered messages will be reported by grid-coordinates, and activity in even numbered messages will be reported by polar-coordinates.

After the enemy activity has ceased, the students will reroll the wire to the CP. A critique is then held at a point where the terrain can be viewed.

SUMMARY:

The problem may be so organized on the terrain that the use of auxiliary OP's to cover defilade areas in an assigned sector can be brought out. The problem should be preceded by a lecture and discussion on the organization of observation in a unit. A discussion of the functioning of the forward artillery observer attached to the infantry may also be planned.

EXERCISE XXV

PURPOSE:

TO INTRODUCE THE STUDENTS TO BASIC SCOUTING AND PATROLLING.

TIME:

10-12 hours (1200-2400)

DESCRIPTION:

The exercise is a two-phase problem consisting of day and night "sneak" reconnaissance patrols.

1. *Day Reconnaissance Patrols:* The day reconnaissance phase should consist of patrol missions through previously selected terrain. Along the general routes the "enemy" should be patrolling or laying ambushes. The patrols should make a written report of all that happened on their mission.

2. *Night Reconnaissance Patrols:* The day phase should be so organized as to allow the night patrols prior daylight observation. After dark the patrol should be sent out into the area on definite missions against a tactically emplaced "enemy"

SUMMARY:

The problem should be designed so that the students learn the basic precepts of both day and night reconnaissance, it being advisable, in some cases, to have the day and night phases on the same terrain so that the differences and difficulties of day and night patrolling can be better stressed.

EXERCISE XXVI

PURPOSE:

1. TO PRACTICE OBSERVATION AND RECONNAISSANCE.

TIME:

10-12 hours.

DESCRIPTION:

This problem has two phases:

1. *The OP Phase:* The patrol should establish its OP early in the day to observe the "enemy", recording and reporting all enemy activities. While operating the OP, the patrol leaders may send out scouts to reconnoiter blind spots and verify routes for the night patrol. Aerial photos can be used in the terrain study.

2. *The Night Patrol:* This phase is similar to the second phase of Exercise XXV, but the patrols must confirm their daylight observations and investigate the OP's blind spots.

SUMMARY:

The patrol must keep local OP security in accordance with a blind situation and exercise all functions of a sneak reconnaissance patrol in enemy territory.

EXERCISE XXVII

PURPOSE:

TO PRACTICE THE INSTALLATION AND

MANNING OF AN OP UNDER COVER OF DARKNESS FOR DAYLIGHT OBSERVATION.

TIME REQUIRED:
0300-1100

DESCRIPTION:

Students should be given definite sectors in which they are to establish their patrol OP's. Their OP's should be established before daylight and after dawn must be operated without any changes in the camouflage or position. A demonstration usually of the breaking of bivouac and the moving out of the area by the "enemy" should begin after daybreak. In addition to this there should be a series of planned incidents to test the students' powers of observation and their ability to locate ground positions on their overlays, etc.

This exercise may be run with many variations but the use of darkness for the picking of OP positions, digging them in if it is a stabilized situation, and the use of camouflage, etc., must be brought out. OP sites may be picked from maps or aerial photos (floating line) and established at night without any prior ground reconnaissance.

EXERCISE XXVIII

PURPOSE:

(1) TO TEACH THE MECHANICS OF A COMBAT-RECONNAISSANCE PATROL.
(2) TO GIVE PRACTICE IN MOVEMENT, CONTROL AND MANEUVER.

DESCRIPTION:

Students are formed into a combat reconnaissance patrol. Instructor will assign each man a job, appoint a patrol leader and will then cover all points in briefing and preparation. A situation will be given. The patrol will then move out on its mission with an instructor with both forward and rear elements. Correct movement will be stressed and control will be checked and corrected when errors occur.

The students will carry all the proper armament and communication equipment. Along the selected route a previously prepared ambush will be laid. It will open fire (blanks) on the forward element when it is committed to the area. The instructor will halt the action and movement of the patrol when the patrol is fired upon. He will arbitrarily designate certain casualties and ask the patrol leader for his decision and action to be taken. The patrol leader should give his decision and then the entire group will be assembled to listen to a critique on the decision. A general discussion led by the instructor will then take place and the approved solution based on the application of small unit tactics to the terrain will be given. The rear element can then replace the forward element and the patrol will proceed along further to another previously prepared ambush where the same procedure will be repeated.

SUMMARY:

This exercise must be carefully planned and its success will depend on the proficiency of the instructors. It should be followed by an exercise that will be completely tactical, which should include time and space factors, preparation, briefing, patrol action, and interrogation of return. In the last tactical exercise the enemy action can be the same and a decision from the patrol leader on his action when fired upon can be similar to the practice exercise. Smoke grenades should be used to cover withdrawals. A critique should be held at the end of the exercise.

EXERCISE XXIX

PURPOSE:

TO PRACTICE PATROLLING AGAINST A HEAVILY ORGANIZED DEFENSIVE POSITION IN A STABILIZED SITUATION.

DESCRIPTION:

The area selected should be tactically sound and preferably should be covered with thick natural vegetation, particularly if a jungle situation is to be simulated. Mutually supporting pill boxes, complete with connection trenches, sniper and troop dispositions and minefield protection, must be established by the "enemy".

Student patrols operating against the defense may be either sneak or combat reconnaissance type depending upon given situation. Ample time must be provided for briefing and a detailed terrain study by the participating personnel. O.P.'s, maps, aerial photos, sand tables, etc., can be utilized during the preparation phase. Rehearsals may be conducted. Definite missions, primary and secondary, must be assigned along with specific zones of patrol operation. Sound power phones or radios in communication with the CP can be issued to maintain control and the necessary coordination among all participating patrols. A complete friendly situation including outguards, minefields, wire, use of passwords etc., should be incorporated to add realism.

SUMMARY:

This exercise may be run with many variations and with emphasis placed on all or any of the planning, briefing, operational or interrogational phases. The use of umpires placed in enemy installations to determine possible casualties and student proficiency will add greatly to the success of the problem.

This type of a problem is often neglected in training programs due to the amount of preparation involved. However, a great majority of actual combat patrolling will actually be against such defensive installations. If the time, equipment, and personnel are available, it should never be omitted in a training program. Regardless of local training limitations, small scale problems of this nature can usually be organized.

EXERCISE XXX

PURPOSE:

TO GIVE THE STUDENTS PRACTICE IN FOLLOWING A MOVING SITUATION BY PATROLLING AND OBSERVATION.

TIME:

Three days

DESCRIPTION:

This exercise should incorporate the former problems in a longer and more detailed situation (a three-day problem). The problem should be set up to cover several phases and the students will be given areas to cover, and they will proceed either mounted or on foot to determine the areas occupied by the enemy. The first phase, then, will be *contact,* during which the OP's will be selected and the enemy is located. The second phase, to cover the first night and second day, will be done from the OP. By this time enemy organization should be recognized and the positions generally known. Scouts and patrols should be used. During the third phase patrols should be sent out on the basis of specific S-2 missions as result of the OP work.

The enemy should execute various maneuvers against the students: they should attempt flanking attacks, retreat, send small raiding parties out after the OP's, and they should use all other enemy security measures. This problem should be very carefully controlled and timed; the degree of success of this problem will depend on the size of the unit operating as an enemy and on the amount of instructors available for control.

TRAINING AIDS

Aside from the use of simulated "enemy" opposition, which is the best of all possible training aids in scouting and patrolling, numerous mechanical and written aids may be devised and used. Enlarged maps, photo mosaics, sketches and sand tables should be used by training officers when conducting the briefing and interrogation phases of the program. Large charts which may be copied from the following sample forms are also very useful.

The training officer should endeavor to see that all equipment issued to our troops and used in active theatres that has a relationship to reconnaissance is available. American ordnance, artillery, vehicles, and armor is of particular value during observation training. Enemy equipment of all types is very valuable for training, particularly after training in the identification and recognition phases of American equipment has been undertaken. Many times local improvisations, mounted on American vehicles and weapons, can be used to simulate "enemy" equipment. For example, jeeps have been used as a base for construction of plywood "enemy" armored vehicles with good results.

MEANS OF CONTROL

In conducting problems against the opposition, radios such as the SCR 536, SCR 300, SCR 511 and SCR 510, are very useful. "Enemy" patrols and installations can be connected with the student C.P. by means of these sets. In this manner a verbal picture of the activities of the student patrols can be relayed to the training officer. Zones and check points can be established in the "enemy" sector so that positive identifications of patrols can be made, as they should have already designated their patrol routes prior to departure on their mission.

On some occasions skilled men with radios can be designated to follow specified patrols at a distance and give a running account of the patrol conduct to the training officer. Sound power phones may also be used in a lesser degree for the same purpose.

Any means of control and checking that can be devised to let the training officer know of patrol performance should be used. Care must be taken, however, to prevent these means of control from becoming a hindrance to the patrol's performance of the mission. For this reason the radio net check idea used by the enemy is desirable as actual contact with the patrol is omitted. In some problems students can be issued chalk with instructions to make designated marks (and time) on certain objects in the area of their mission or objective. This means of control helps place responsibility on the individual and is an aid in testing and confirming the student's proficiency against "enemy" installations.

TESTS

Written tests should be conducted in scouting and patrolling but care should be taken to avoid placing too much graded emphasis on them. A grade in this subject should largely be determined by observation of the individual in the field. A high grade in the written test should not be considered an infallible indication of actual scouting and patrolling proficiency.

A written test, if used, should consist mainly of

practical questions and should not be made up of too many questions of the "by the numbers", stereotyped kind. For example, a question of this type is much better:

"Your combat reconnaissance patrol of thirty-eight men has just come through a fire fight and you have succeeded in neutralizing the opposition on what you estimate is the enemy OPL. You have four PW, eight walking casualties (one of whom is the squad leader of your number one squad), and two casualties who cannot walk. You have not accomplished your mission. What orders will you give?"